AF769382

PETROLEUM SCIENCE AND TECHNOLOGY

Additional books in this series can be found on Nova's website under the Series tab.

Additional e-books in this series can be found on Nova's website under the e-book tab.

PETROLEUM SCIENCE AND TECHNOLOGY

ALTERNATIVE JET FUELS

DEVELOPMENT AND VIABILITY

PETROLEUM SCIENCE AND TECHNOLOGY

ALTERNATIVE JET FUELS

DEVELOPMENT AND VIABILITY

JENNA L. AIKEN
EDITOR

nova publishers
New York

Copyright © 2014 by Nova Science Publishers, Inc.

All rights reserved. No part of this book may be reproduced, stored in a retrieval system or transmitted in any form or by any means: electronic, electrostatic, magnetic, tape, mechanical photocopying, recording or otherwise without the written permission of the Publisher.

For permission to use material from this book please contact us:
Telephone 631-231-7269; Fax 631-231-8175
Web Site: http://www.novapublishers.com

NOTICE TO THE READER

The Publisher has taken reasonable care in the preparation of this book, but makes no expressed or implied warranty of any kind and assumes no responsibility for any errors or omissions. No liability is assumed for incidental or consequential damages in connection with or arising out of information contained in this book. The Publisher shall not be liable for any special, consequential, or exemplary damages resulting, in whole or in part, from the readers' use of, or reliance upon, this material. Any parts of this book based on government reports are so indicated and copyright is claimed for those parts to the extent applicable to compilations of such works.

Independent verification should be sought for any data, advice or recommendations contained in this book. In addition, no responsibility is assumed by the publisher for any injury and/or damage to persons or property arising from any methods, products, instructions, ideas or otherwise contained in this publication.

This publication is designed to provide accurate and authoritative information with regard to the subject matter covered herein. It is sold with the clear understanding that the Publisher is not engaged in rendering legal or any other professional services. If legal or any other expert assistance is required, the services of a competent person should be sought. FROM A DECLARATION OF PARTICIPANTS JOINTLY ADOPTED BY A COMMITTEE OF THE AMERICAN BAR ASSOCIATION AND A COMMITTEE OF PUBLISHERS.

Additional color graphics may be available in the e-book version of this book.

Library of Congress Cataloging-in-Publication Data

ISBN: 978-1-63321-626-6

Published by Nova Science Publishers, Inc. † New York

CONTENTS

PREFACE

The federal government has encouraged the development and use of alternative fuels to reduce greenhouse gas emissions associated with aviation and to enhance economic development and energy security for the United States. To help achieve these goals of reducing greenhouse gas emissions, the aviation industry is actively supporting alternative jet fuels. This book examines the role of the federal government in the development and use of alternative jet fuels; and key challenges to developing and using alternative jet fuels and actions that the federal government plans to or could take to help address those challenges.

Chapter 1 - The federal government has encouraged the development and use of alternative fuels to reduce greenhouse gas emissions associated with aviation and to enhance economic development and energy security for the United States. To help achieve these goals of reducing greenhouse gas emissions, the aviation industry is actively supporting alternative jet fuels.

GAO was asked to provide information on the progress and challenges to developing and using alternative jet fuels in the United States. This report examines (1) the role of the federal government in the development and use of alternative jet fuels and (2) key challenges to developing and using alternative jet fuels and actions that the federal government plans to or could take to help address those challenges.

GAO interviewed officials from five federal agencies—FAA, USDA, DOE, DOD, and EPA. GAO selected these agencies for review because GAO identified them as the federal agencies most involved in the development and use of alternative jet fuels. GAO also reviewed relevant literature and federal and industry documents and discussed challenges and potential federal actions with 23 stakeholders from government, academia, and the private sector, selected to represent a range of perspectives and expertise in areas related to each step in the development and use of alternative jet fuels. GAO is not making recommendations in this report. DOT, USDA, DOE, DOD, and EPA reviewed a draft of this report and provided technical comments that were incorporated as appropriate.

Chapter 2 - This study summarizes the best available public data on the production, capacity, cost, market demand, and feedstock availability for the production of biomass-based diesel and jet fuel. It includes an overview of the current conversion processes and current state-of-development for the production of biomass-based jet and diesel fuel, as well as the key companies pursuing this effort. The discussion analyzes all this information in the context of meeting the Renewable Fuels Standard mandate, highlights uncertainties for the future industry development, and key business opportunities.

In: Alternative Jet Fuels
Editor: Jenna L. Aiken
ISBN: 978-1-63321-626-6
© 2014 Nova Science Publishers, Inc.

Chapter 1

ALTERNATIVE JET FUELS: FEDERAL ACTIVITIES SUPPORT DEVELOPMENT AND USAGE, BUT LONG-TERM COMMERCIAL VIABILITY HINGES ON MARKET FACTORS*

United States Government Accountability Office

WHY GAO DID THIS STUDY

The federal government has encouraged the development and use of alternative fuels to reduce greenhouse gas emissions associated with aviation and to enhance economic development and energy security for the United States. To help achieve these goals of reducing greenhouse gas emissions, the aviation industry is actively supporting alternative jet fuels.

GAO was asked to provide information on the progress and challenges to developing and using alternative jet fuels in the United States. This report examines (1) the role of the federal government in the development and use of alternative jet fuels and (2) key challenges to developing and using alternative jet fuels and actions that the federal government plans to or could take to help address those challenges.

---

* This is an edited, reformatted and augmented version of a United States Government Accountability Office publication, No. GAO-14-407, dated May 2014.

GAO interviewed officials from five federal agencies—FAA, USDA, DOE, DOD, and EPA. GAO selected these agencies for review because GAO identified them as the federal agencies most involved in the development and use of alternative jet fuels. GAO also reviewed relevant literature and federal and industry documents and discussed challenges and potential federal actions with 23 stakeholders from government, academia, and the private sector, selected to represent a range of perspectives and expertise in areas related to each step in the development and use of alternative jet fuels. GAO is not making recommendations in this report. DOT, USDA, DOE, DOD, and EPA reviewed a draft of this report and provided technical comments that were incorporated as appropriate.

What GAO Found

The federal government supports the development and use of alternative jet fuels through both broad and targeted initiatives. Broad national strategies promote the development of a variety of alternative fuels—including alternative jet fuel—to help achieve national goals, such as securing energy independence, fostering economic development, and reducing greenhouse gas emissions. In addition, the renewable fuel program—established by law in 2005 to encourage greater use of renewable fuels and administered by the Environmental Protection Agency (EPA)—requires that U.S. transportation fuels contain certain amounts of renewable fuels annually, increasing from 9-billion gallons in 2008 to 36-billion gallons in 2022. The other four federal agencies that GAO reviewed— Department of Transportation's (DOT) Federal Aviation Administration (FAA), Department of Agriculture (USDA), Department of Energy (DOE), and Department of Defense (DOD)—directly support alternative jet fuels through targeted goals, initiatives, and interagency and industry coordination efforts. For example, FAA set a goal for the U.S. aviation industry to use 1-billion gallons of alternative jet fuels annually by 2018. The four agencies also sponsor research that specifically targets alternative jet-fuel development or provide direct support for its future commercial production, or both. For example, FAA and DOD support research to determine the technical feasibility of using new alternative jet fuels on aircraft and in existing infrastructure. Also, USDA, DOE, and DOD have coordinated their activities to support the future construction or retrofit of multiple domestic commercial- or pre-commercial-scale production facilities to produce alternative fuels, including alternative jet fuels. Specifically, in

May and June 2013, four private fuel producers received awards totaling $20.5 million in federal funds, with private industry paying at least 50 percent of the cost.

Achieving price competitiveness for alternative jet fuels is the overarching challenge to developing a viable market. No alternative jet fuels are currently commercially available at prices competitive with conventional jet fuels. The 23 stakeholders that GAO interviewed most frequently cited high development costs and the uncertainty of federal regulations and policies as primary reasons why alternative jet fuels are not priced competitively and believe that federal activities are needed to help advance the alternative jet-fuels industry. For example, according to 10 stakeholders, fuel producers face difficulties in obtaining private investment to help construct commercial-scale fuel production facilities, in part because of concerns about the supply and high cost of feedstock (the source used to produce the fuel, such as crops) and high capital costs. Also, 13 stakeholders stated that continued uncertainty about the future of current federal policies—particularly the renewable fuel program—generally causes potential investors to discount the value of federal subsidies, discounting that, in turn, limits the support these policies may provide the industry. Stakeholders identified a variety of federal actions to advance alternative jet-fuels development, including continuing current federal research efforts, providing greater regulatory and policy certainty, and giving more direct financial support. However, even if the cost to produce alternative jet fuels is reduced, market factors may still determine the long-term success of the industry. The main market factors identified by stakeholders were (1) comparative value of competing end products, (2) feedstock prices, and (3) the costs of conventional jet fuels.

Abbreviations

BBEDCA	Balanced Budget and Emergency Deficit Control Act
BETO	Bioenergy Technologies Office
CAAFI	Commercial Aviation Alternative Fuels Initiative
CLEEN	Continuous Lower Energy, Emissions and Noise
COE	Center of Excellence
DOD	Department of Defense
DOE	Department of Energy
DOT	Department of Transportation
EPA	Environmental Protection Agency

EISA	Energy Independence and Security Act
EPA	Environmental Protection Agency
FAA	Federal Aviation Administration
HEFA	Hydroprocessed Esters and Fatty Acids
USDA	Department of Agriculture
MOU	memorandum of understanding
NASA	National Aeronautics and Space Administration
NextGen	Next Generation Air Transportation System
R&D	research and development
RFS	Renewable Fuel Standard
RIN	renewable identification number

May 7, 2014

Congressional Requesters

Governments and the international aviation community are seeking to limit aviation greenhouse-gas emissions, which according to many scientists and the Intergovernmental Panel on Climate Change,[1] negatively affect the earth's climate. While aviation emissions represent a small share of total U.S. carbon dioxide emissions—the most significant greenhouse gas emission emitted by aviation[2]—we have found that the aviation sector's share is estimated to increase at a greater rate than other sources of fossil fuel combustion, in part due to projected growth in the aviation industry and the fact that unlike other transportation modes, the aviation industry does not have near-term alternatives to liquid fuels.[3] The development and use of alternative jet fuels, particularly fuels derived from renewable sources (renewable fuels), have the potential to reduce greenhouse gas emissions associated with aviation, in part because crops used to produce these fuels, called feedstock, remove carbon dioxide from the air as they grow.[4] In addition to potentially reducing greenhouse gas emissions associated with aviation, developing dependable domestic sources of alternative fuels could help enhance energy security for the United States. According to the Energy Information Administration, jet fuel prices more than tripled from 2002 to 2012.[5] Over this time period, jet fuel's share of airlines' total operating costs increased from 13 percent to 31 percent, making fuel the largest component of operating costs in 2012, based on analysis of airlines' financial reports to the Department of Transportation (DOT). As a result, the cost of producing alternative jet fuels is an important factor in the extent to which they provide a viable option for the aviation sector to replace more conventional petroleum-based jet fuel.

To encourage greater use of alternative fuels for all sectors, the federal government has taken steps to encourage the development of a commercially viable, alternative-fuels industry.[6] In 2005, Congress, through the Energy Policy Act of 2005,[7] established the renewable fuels program, establishing the basis for the Administrator of the Environmental Protection Agency (EPA) to adopt standards through regulations, which for the first time require that surface transportation fuels sold in the United States contain a minimum amount of renewable fuels.[8] In March 2011, the White House put forward a plan to produce more domestic oil and reduce U.S. dependence on foreign oil by leveraging alternative fuels and increasing fuel efficiency. In June 2013, the President released a Climate Action Plan stating that renewable alternative fuels play an important role in reducing greenhouse gas emissions from the transportation sector, increasing energy security, and fostering rural economic development. The plan proposed increased funding for research, development, and deployment of advanced biofuels.[9] The aviation industry also has efforts under way, including partnerships with fuel producers and other stakeholders, to enhance the development and use of alternative jet fuel to help achieve the industry's goals of reducing greenhouse gas emissions.[10] Currently, some types of alternative jet fuels are approved for use in civilian and military aircraft.

You asked us to provide information on the progress in and challenges to developing and using alternative jet fuels in the United States. This report examines (1) the role of the federal government in the development and use of alternative jet fuels and (2) key challenges to developing and using alternative jet fuels and actions that the federal government plans to or could take to help address those challenges.

To identify the role of the federal government, we interviewed officials and reviewed strategic plans, performance reports, and other documents relevant to alternative jet fuels from five federal agencies: the DOT's Federal Aviation Administration (FAA), the Department of Agriculture (USDA), the Department of Energy (DOE), the Department of Defense (DOD) and its military departments (Army, Navy, and Air Force), and the EPA. We selected these agencies for review because we identified them as the federal agencies most involved in the development or use of alternative jet fuels. We also reviewed White House and other broad or government-wide documents relevant to alternative fuels generally. We reviewed relevant memorandums of understanding or other agreements between the five selected federal agencies and other entities. To identify key challenges and planned or possible actions to address key challenges, we selected and interviewed 23 stakeholders representing government, academia, and the private sector. We selected these stakeholders based on criteria that included type and depth of experience and knowledge in the area of alternative jet fuels, recognition in the professional community, and representation of a range of expertise in areas related to each step in the development and use of alternative jet fuels, such as feedstock development or fuel production. We asked these stakeholders a series of semi-structured, open-ended questions about economic, policy, and technological issues related to developing and using alternative jet fuels. In addition, we interviewed officials from the five selected federal agencies, as well as other non-federal alternative jet-fuels industry stakeholders, including fuel producers, airlines, aircraft manufacturers, and environmental groups. We also reviewed literature on challenges to developing and using alternative jet fuels. We focused our study on federal efforts, except to the extent that international or private sector efforts are coordinated with federal efforts. We did not provide an exhaustive list of federal initiatives related to alternative jet fuels or rank the planned or potential federal actions to address the key challenges that we identified. Appendix I provides a more detailed description of our scope and methodology.

We conducted this performance audit from February 2013 to May 2014 in accordance with generally accepted government auditing standards. Those standards require that we plan and perform the audit to obtain sufficient, appropriate evidence to provide a reasonable basis for our findings and conclusions based on our audit objectives. We believe that the evidence obtained provides a reasonable basis for our findings and conclusions based on our audit objectives.

BACKGROUND

Definitions vary somewhat, but for the purpose of this report, we define alternative jet fuels as drop-in liquid fuels that are derived from non-petroleum feedstocks, including renewable biomass (such as crop and tree residues, algae, or separated municipal solid waste) and some nonrenewable sources (such as natural gas or coal). Because they are "drop-in," alternative jet fuels can replace conventional petroleum-based jet fuel (i.e., conventional jet fuel) without the need to modify aircraft engines and fuel distribution infrastructure. This definition means that alternative jet fuels are a substitute for conventional jet fuel and would directly compete with it on the commercial market. Most development activities—both federal and private industry activities—are focused on alternative jet fuels derived from renewable sources, in part, because of the prominence of environmental sustainability in national strategies and a federal requirement that any alternative jet fuel for operational use procured by any federal agency must have lifecycle greenhouse gas emissions less than or equal to conventional fuel of the same purpose.[11]

Alternative jet fuels are generally derived from the same feedstocks and processes as other advanced fuels (e.g., renewable diesel) and even some industrial products (e.g., plastics). As a result, alternative jet fuel producers have some flexibility in producing a variety of fuels and other products, seeking to maximize their profit relative to demand and costs. Figure 1 describes the six segments of the supply chain for the development and use of alternative jet fuels.

The first three segments of the alternative jet-fuel supply chain—feedstock production, feedstock logistics, and fuel production—are common to all alternative fuels, and the first two segments are generally independent of the end-use product or co-products that are produced. For example, the camelina crop can be processed, converted, and refined into multiple end-use products, including renewable gasoline, diesel, and jet fuel, with the relative quantities of each of these end-use products determined by the fuel refiner. The activities associated with producing the camelina (e.g., growing the crop) and the logistical activities of collecting, storing, and transporting the camelina feedstock before it is converted to fuel are the same regardless of the relative quantities of renewable diesel or jet fuel produced. In the third segment of the supply chain, alternative jet fuels are generally co-produced with other end-use products, although fuel producers do have some flexibility in the mix of products they make.

The remaining three segments of the supply chain—fuel testing and approval, fuel distribution, and end use—include activities that are specific to jet fuel. For example, this supply chain includes a segment for fuel testing and approval because before any alternative jet fuel can be approved for commercial or military use, it must meet unique safety and performance standards that are more rigorous than standards for other alternative transportation fuels. Standards for alternative jet fuels are set out in applicable standards controlled by ASTM International[12] and the appropriate military department within DOD (the Navy and the Air Force).[13]

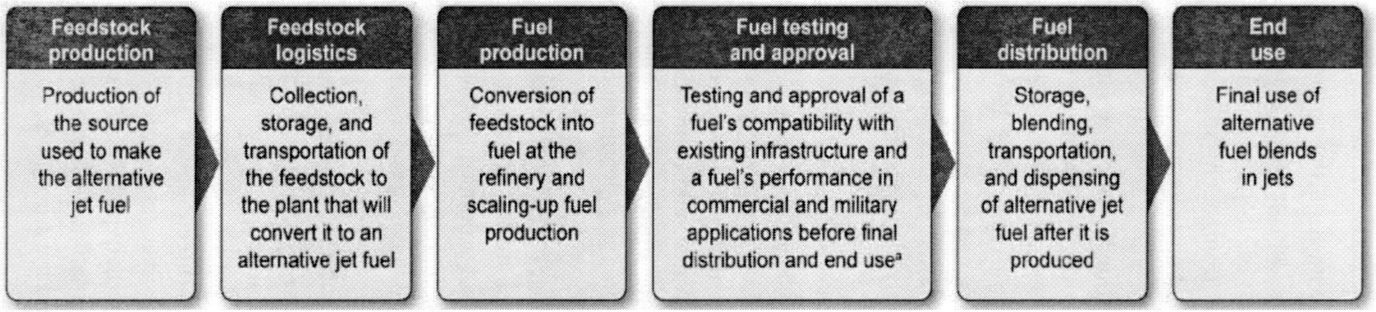

Source: GAO.

Note: The alternative jet fuel supply chain and segments were developed through GAO review and analysis of documents from federal agencies and industry with input from officials of each federal agency and department that we interviewed.

[a]Fuel testing and approval activities may occur at various stages of the supply chain. They may also include environmental testing, such as tests to evaluate fuel-cycle energy use and lifecycle greenhouse gas emissions directly associated with producing an alternative jet fuel.

Figure 1. Alternative Jet Fuel Supply Chain.

The requirements for fuel testing and approval under ASTM International and military standards vary by the characteristics of the fuel and the feedstock and production process used, but generally the requirements include a significant amount of fuel, engine, and aircraft testing. Initially, a fuel may be tested in a laboratory using small quantities of fuel (as little as 500 milliliters), but as it progresses through the approval process, fuel quantity requirements could reach as much as 225,000 gallons if extensive engine testing is required.

The last two segments of the alternative jet fuel supply chain—fuel distribution and end use—also include activities specific to jet fuels, but not specific to alternative jet fuels relative to conventional jet fuel. Specifically, according to the applicable ASTM International and military standards, once an alternative jet fuel is produced, certified, and released under the applicable standard, it also meets the standards for conventional jet fuel. Accordingly, alternative jet fuels can be seamlessly integrated into the existing jet fuel distribution system and onto aircraft without changes to any infrastructure.

Although no alternative jet fuels are currently available in the United States at a competitive price, as discussed later in this report, two alternative jet fuel production processes—referred to as the FischerTropsch process[14] and the HEFA process[15]—are approved for use by commercial and military aviation. Under the previously mentioned ASTM International and military standards, fuels produced through these two processes are approved for up to a 50 percent blend with conventional jet fuel.

As with conventional jet fuel production processes, both of these processes produce multiple end-use products, including, for example, diesel and jet fuel.

The fuel producer, to a limited extent, determines the relative quantities of jet fuel, diesel, and other products to be produced. For many feedstocks used in the HEFA process, changing the product ratio to produce more jet fuel and less diesel, however, is more costly because it requires additional processing and increases the proportion of output that is comprised of less-valuable co-products, such as liquefied petroleum gas. Seven other production processes for alternative jet fuels, such as converting alcohols to jet fuel, are undergoing review in the ASTM International testing and approval process and in some military departments' testing and approval processes.

Both Broad and Targeted Federal Alternative-Fuels Initiatives Support the Development and Use of Alternative Jet Fuels

Broad Federal Strategies and Initiatives Have Supported Alternative Jet-Fuel Development and Use

The White House has developed broad national strategies that promote the development of alternative fuels to help secure energy independence, foster economic development, and reduce greenhouse gas emissions. For example, the *Blueprint for a Secure Energy Future* (March 2011), the *National Bioeconomy Blueprint* (April 2012), and the *President's Climate Action Plan* (June 2013), all describe how supporting the development of alternative fuels can contribute to achieving these broad national goals. However, these strategies also note that alternative fuels are only part of a wide variety of other complementary activities working toward the same goals. For example, according to the President's Climate Action Plan, meeting U.S. greenhouse gas-emission reduction goals[16] depends not only upon the development and use of alternative fuels, but also on numerous other activities such as increasing fuel economy standards, expanding and modernizing the electric grid, and improving the energy efficiency of homes and businesses.

Some initiatives led by one or more of our five selected federal agencies involved in the development or use of alternative jet fuel[17] help support these broad national strategies and are often similarly broad—focusing on issues that are common to a variety of alternative fuels, including alternative jet fuels. For example, to support the development of technologies and processes necessary for the commercial production of biofuels at prices that are competitive with conventional fuels, USDA and DOE jointly administer the Biomass Research and Development Initiative, which assists in developing these technologies through research, development, and demonstration projects. According to USDA and DOE officials, most of the initiative's projects do not target alternative jet fuels specifically. However, officials from these agencies noted that general scientific advancements that broad initiatives identify can also advance alternative jet fuels development specifically. For example, research that improves the efficiency of a particular fuel-conversion process often supports the development of all fuels that the process could produce.

In 2007, the Energy Independence and Security Act (EISA) of 2007 expanded the Renewable Fuel Standard (RFS) to cover most surface transportation fuels, such as fuels for use in motor vehicles and engines and nonroad vehicles and engines, but not jet fuel.[18] Overall, jet fuel is a fraction of the total transportation fuel consumed in the United States.[19] The expanded RFS generally required that covered transportation fuels contain 9-billion gallons of renewable fuels in 2008, with renewable fuels' volumes increasing annually to 36-billion gallons in 2022. To demonstrate compliance with the RFS, fuel producers or importers use renewable identification numbers (RINs). Fuel producers or importers can obtain RINs by purchasing and blending renewable fuels themselves, or they can purchase RINs from renewable fuel producers, importers, blenders, or other parties. In this way, the renewable fuel program has created a market for RIN credits. While jet fuel is not used to calculate a fuel producer's or importer's renewable fuel obligation, EPA determined in its March 2010 final rule for the expanded renewable fuel program that some feedstocks and conversion processes for renewable jet fuel qualify as "advanced biofuels," one of the new categories of renewable fuel established by EISA.[20] In addition, through regulations issued in March 2013, EPA clarified that some renewable diesel processes that had been previously evaluated included jet fuel and also approved additional jet-fuel pathways.[21] If alternative jet fuels produced through these approved processes generate RIN credits—which the fuel producer may then sell to others—their sale may help subsidize the cost of producing qualifying alternative jet fuels.

Four Federal Agencies Support Alternative Jet Fuels Development and Use through Targeted Goals, Initiatives, and Coordination Efforts

Four of the selected agencies—FAA, DOD, USDA, and DOE—support initiatives that target alternative jet-fuel development or use specifically. FAA and DOD have established specific goals for using alternative jet fuels in commercial and military aircraft and support research and development (R&D) activities—such as testing to approve new alternative jet fuels—to help them to achieve these goals. DOD, USDA, and DOE also have initiatives that provide direct financial support for future alternative jet-fuel production on a commercial-scale. All of these agencies coordinate their alternative jet fuel-related efforts with industry and other stakeholders through partnerships and agreements.

Federal Agencies' Usage Goals

FAA and DOD have established usage goals specifically for alternative jet fuels. In fiscal year 2012, FAA set a goal for the U.S. aviation industry (including commercial and military aircraft) to use 1-billion gallons of alternative jet fuels annually by 2018 with the intent of encouraging commercial production.[22] According to FAA officials, this represents about 5 percent of the predicted jet fuel consumption for domestic airlines and the military in 2018. Achieving this goal, however, will depend on a variety of factors, including support from other federal agencies and industry stakeholders. USDA and other nongovernmental stakeholders have stated their intent to help enable commercial production of alternative jet fuels in support of FAA's goal through existing programs and expanded collaboration.

Two of DOD's military departments—the Navy and the Air Force—have also established usage goals for alternative fuels, including alternative jet fuels.[23] To support these usage goals, the Navy and Air Force are willing to purchase alternative fuels that meet specific criteria, including availability at a price that is competitive with conventional fuels.

- The Navy's 2010 *A Navy Energy Vision for the 21st Century*[24] states that increasing its use of alternative energy—including alternative jet fuels—will help protect it from energy price volatility and supply disruptions. The plan sets a goal of deriving 50 percent of total Navy energy consumption afloat[25]—including its jet fuel consumption—from alternative sources by 2020, which, according to Navy estimates, would require using about 336-million gallons of alternative fuels annually (both marine and jet fuels) by 2020.[26] The Navy consumes over 600-million gallons of petroleum-based aviation fuel each year, which according to a Navy official, constitutes about 40 percent of its total petroleum consumption. In addition to setting quantitative goals, the plan established a goal of demonstrating (which the Navy completed in July 2012) and deploying the Great Green Fleet—a group of ships and aircraft fueled by alternative jet fuels and other alternative energy sources—by 2016.
- The Air Force's 2013 *U.S. Air Force Energy Strategic Plan*[27] includes a goal of increasing the use of cost-competitive drop-in alternative jet-fuel blends for non-contingency operations to 50 percent of total consumption by 2025. According to the plan, the Air Force consumes about 2.5 billion gallons of jet fuel each year, accounting for about 80 percent of its total energy consumption. The plan states that using

alternative jet fuels could help to diversify the types and secure the quantities of energy that are needed to perform the Air Force's missions, which are currently "heavily dependent" upon petroleum and petroleum-derived fuels, posing significant strategic and security vulnerabilities on energy supplies.

Federal Agency Initiatives for R&D and Direct Financial Support

Two types of federal initiatives specifically support the development and use of alternative jet fuels—sponsoring R&D and direct financial support for future commercial production. FAA, DOD, and DOE support R&D activities to target alternative jet fuels specifically, such as testing to approve new alternative jet fuels or research to determine the environmental or economic impact of using them. USDA, DOE, and DOD have also taken some steps to provide direct financial support for future commercial-scale production of alternative jet fuels.[28]

Research and Development Activities

DOT—primarily FAA—supports activities to determine the technical feasibility and impact of using alternative jet fuels, including ways to reduce the cost of production, through FAA's Continuous Lower Energy, Emissions and Noise (CLEEN) program and Centers of Excellence (COEs), and through other DOT activities.

- **CLEEN Program:** Launched by FAA's Office of Environment and Energy in 2010, CLEEN is a cost-sharing program that, among other things, supports fuel-testing activities to generate data that can be used to support the approval of new alternative jet fuels.[29] According to FAA officials, under the program, FAA is providing about $125 million in matching funds (of which about $93 million was provided through fiscal year 2013) to five projects with engine and airframe manufacturers. According to FAA officials, four of these projects address issues related to alternative jet fuels development or use. For example, through fiscal year 2013, FAA has awarded about $5.5 million through the CLEEN program to conduct laboratory and engine-component tests of advanced alternative jet fuels that could be approved for commercial use by ASTM International. FAA has announced its plans to implement a follow-on program in 2015, called CLEEN II—when FAA plans to end the initial CLEEN program. Like the initial program, CLEEN II's goals will include developing and demonstrating drop-in sustainable alternative jet fuels. However, FAA officials told us that under CLEEN II they intend to place an emphasis on advancing coordinated test methods and capabilities to reduce testing cost and time and the possibility of redundant testing by multiple engine manufacturers.

- **Centers of Excellence:** FAA also sponsors research studies related to the environmental impact of using alternative jet fuels through its COEs. Beginning in 2003, FAA sponsored the Center of Excellence for Aircraft Noise and Aviation Emissions Mitigation, named Partnership for AiR Transportation Noise and Emissions Reduction (PARTNER), a collaborative effort that researched solutions for existing and anticipated aviation-related noise and emissions problems.[30] According to FAA officials, five PARTNER projects focused on alternative jet fuels specifically. For example, one project studied the economic feasibility, production potential, and environmental impact of alternative jet-fuel use. Upon the expiration of PARTNER's 10-year cooperative agreement in September 2013 and as required in the FAA Modernization and Reform Act of 2012,[31] FAA selected a team of universities to form a new COE for Alternative Jet Fuel and Environment, named the Aviation Sustainability Center (ASCENT), with research goals that include better understanding ways to reduce the costs of production processes and ways to meet FAA's goal of using 1-billion gallons of alternative jet fuels by 2018. ASCENT is being led by Washington State University and the Massachusetts Institute of Technology, and is expected to receive at least $4 million annually for 10 years to explore ways to meet FAA's environmental and energy goals, including sustainable alternative jet fuels.
- **DOT activities:** DOT also helps fund research through broad agency announcements for FAA-sponsored projects that address specific alternative jet fuels' testing needs. Specifically, in 2010 DOT invited research in four priority areas: development of novel "drop-in" alternative jet fuels, alternative jet fuels' quality control, sustainability guidance for alternative jet fuels' users and performance, and durability testing of new fuels. According to FAA officials, DOT's John A. Volpe National Transportation Systems Center has administered six FAA-sponsored broad agency-announcement research projects related to alternative jet fuels at a total cost of about $7 million. For example, DOT provided funds to three alternative jet-

> fuel producers to develop and optimize conversion processes for alternative jet fuels. These funds better positioned the fuel producers to produce fuel for testing purposes and develop knowledge to help overcome key technical hurdles to commercial-scale production.

DOD also supports activities to test and approve alternative jet fuels. All three DOD military departments coordinate their alternative jet-fuel testing and approval efforts through the Tri-Service Alternative Fuels Working Group. According to Air Force officials, this working group helps share data, reports, testing, and certification practices across DOD and is working toward developing a department-wide certification strategy. Specifically, the Army, Navy, and Air Force test alternative jet fuels to ensure that they are safe to use on military ships, aircraft, and fuel distribution systems. Their testing programs capture technical data through laboratory, component, engine, fuel system, and weapon system tests that evaluate the effects of changes in fuel chemistry and properties on the performance and reliability of military equipment. According to DOD officials, the department purchased about 1.5-million gallons of alternative jet fuels to conduct the department's testing and approval activities from fiscal years 2007 to 2013 at a total cost of almost $40 million.[32] Officials from the Navy and Air Force told us that these activities will help enable them to achieve their stated goals for alternative jet-fuels use. DOE also, on behalf of DOD, recently solicited applications for R&D projects that help enable conventional coal-to-liquid production plants to produce commercially viable quantities of jet fuel that have equal or lower greenhouse gas emissions and make significant progress toward being cost-competitive to conventional jet fuel. DOE expects to select and award applications by the end of August 2014, with about $20 million available under the solicitation.[33]

In addition, federal agency officials representing eight federal agencies recently formed an interagency working group that is currently drafting a national R&D strategy for alternative jet fuels. According to members of the working group, a national R&D strategy is needed to create a national vision for alternative jet fuels specifically and a unified federal government approach to help facilitate interaction with external stakeholders, such as industry and academia. As part of the working group's efforts, in January 2014, the working group sponsored a workshop attended by government and industry stakeholders. The workshop identified a variety of challenges to making alternative jet fuels, including challenges associated with feedstock logistics, fuel production and scale-up, fuel certification and qualification, as well as other cross-cutting issues. According to members of the working group, the national strategy for alternative jet fuels will focus on R&D challenges and will not address policy issues.

Direct Financial Support for Future Commercial Production

In June 2011, USDA, DOE, and one of DOD's military departments (the Navy) signed a memorandum of understanding (MOU) that initiated cooperation among these agencies in assisting the development and support of a sustainable commercial biofuels industry, which could produce alternative jet fuels among other types of biofuels. The MOU explained that given the current economic environment, significant startup risks, and competitive barriers of an established conventional fuels market, it is necessary for the federal government to cooperate with private industry to create a strong demand signal and to make targeted investments to achieve the necessary alternative-fuels-production capacity. The stated objective is to construct or retrofit multiple domestic commercial- or pre-commercial-scale advanced drop-in biofuel production facilities.[34] Specific characteristics required for the facilities include that the biofuels that they produce must be capable of meeting military fuel standards at a price that is competitive with conventional jet fuel and have no significant impact on the supply of agricultural commodities for the production of food.

Under the MOU, USDA, DOE, and the Navy stated their intent to contribute $170 million each over 3 years, for an aggregate total of $510 million. Under the authority of the Defense Production Act, Title III,[35] DOE and the Navy planned to fund their share of $340 million for capital investment and production. USDA planned to provide its contribution under the authority of the Commodity Credit Corporation Charter Act.[36] Under this MOU, in June 2012, USDA, DOE, and DOD announced the initiation of and a solicitation for the Advanced Drop-In Biofuels Production Project, which would provide awards for biofuels production facilities over two phases.[37] In May and June 2013, four private companies were selected to receive awards totaling $20.5 million, with private industry paying at least 50 percent of the cost. According to Defense Production Act Title III program officials, the Advanced Drop-In Biofuels Production Project should provide production capacity for about 35 million gallons per year of renewable jet fuels that meet military standards and are available at a price that is competitive to conventional fuels by 2016. But the amount of production capacity is dependent, in part, on the timing and the number of awards for the Advanced Drop-In Biofuels Production Project's second phase. According to DOD officials, the department plans to make its determination for the second phase of awards in July 2014.

More recently, in December 2013, the Secretaries of USDA and the Navy announced another initiative that complements the Advanced Drop-In Biofuels Production Project called Farm to Fleet, which is intended to help the Navy meet its alternative-fuels usage goals. Under the initiative, DOD plans—through its regular domestic bulk-fuel purchases—to issue solicitations in 2014 for the purchase of about 80-million gallons of any combination of jet and marine diesel fuels in 2015 that are blended with at least 10 percent alternative fuels.[38] USDA plans to contribute up to about $161 million (under the authority of the Commodity Credit Corporation Charter Act) toward these fuel purchases to help defray any domestic feedstock costs that would have caused the final alternative fuel to not be price competitive with conventional fuels.

In addition, DOE provides direct financial support for future alternative jet fuels production through its integrated biorefineries program, which was initiated in 2005. Under the program, DOE's Bioenergy Technologies Office (BETO) works in partnership with industry to develop, build, operate, and validate integrated biorefineries at various scales (pilot, demonstration, and commercial). The purpose of these projects is to provide federal support to private industry to help bridge the gap between promising R&D scientific advancements and commercial-scale production by validating fuel conversion technologies at progressively larger scales. According to BETO, federal financial support is essential to help offset the technical and financial risks associated with producing alternative fuels at a commercial-scale. According to BETO officials, DOE has obligated almost $198 million for 14 integrated biorefinery projects related to the development or use of alternative jet fuels.[39] For example, it obligated about $50 million to a fuel producer to demonstrate the technical and economic feasibility of refining algal oil into gasoline, diesel, and jet fuel.

Coordination Efforts with Private Industry and Other Stakeholders

Because of private industry's indispensible role throughout the alternative jet fuel supply chain—such as producing feedstock and fuel—it is critical that the federal government's activities are coordinated with external stakeholders. As a result, USDA, DOE, FAA, and DOD participate in a variety of coordination efforts, such as partnerships with industry and other stakeholders, to identify opportunities to work toward common goals and needs. For example, FAA and other federal agencies participate in the Commercial

Aviation Alternative Fuels Initiative (CAAFI), a public-private partnership[40] formed in 2006 to facilitate the development and deployment of drop-in alternative jet fuels that are intended to reduce all aviation emissions, improve price stability, and support supply security. Key CAAFI efforts have included developing and sharing user guides and tools, as well as organizing workshops for alternative fuel producers and other stakeholders. For example, in December 2013, CAAFI published a user's guide to help alternative jet-fuel producers understand and comply with ASTM International's process to test and approve new alternative jet fuels. The partnership also developed a "Path to Alternative Jet Fuel Readiness" tool that describes the testing and environmental evaluations required to show a new alternative jet fuel's suitability for aviation use and how to best facilitate ASTM International approval. In addition, in January 2013 and January 2014, CAAFI conducted workshops on current regulatory, voluntary, and research efforts related to alternative jet-fuel sustainability issues. Among other things, 2013 workshop participants identified the need to understand and reconcile differences among various approaches to calculating life-cycle greenhouse gas emissions that result from producing alternative jet fuels, and 2014 participants began assessing the differences.

USDA coordinates with private industry and other governmental stakeholders through the FARM to FLY initiative, which was initially established in July 2010 to accelerate the availability of a commercially viable and sustainable domestic alternative jet-fuels industry, increase domestic energy security, establish regional supply chains, and support rural development. USDA expanded the initiative by signing a 5-year FARM to FLY 2.0 resolution with CAAFI, FAA, Airlines for America,[41] and others. Under the expanded resolution, participants agreed to designate personnel for a working group tasked with assessing and proposing ways to support FAA's goal of using 1-billion gallons of alternative jet fuels by 2018. The working group plans to issue a final report by the end of 2018.

In addition to the two efforts described above, federal agencies participate in a variety of other coordination efforts, including the following.

- USDA, DOE, FAA, and DOD work with industry and other stakeholders through regional initiatives that are aimed at advancing alternative fuels within specific regions of the United States.[42]
- DOE's BETO co-sponsored a September 2013 workshop to obtain input from industry, academia, and other experts on optimizing and

integrating the use of natural gas and biomass to produce liquid transportation fuels, including alternative jet fuels.

- FAA has signed onto or agreed to engage in activities under international cooperative agreements with four countries: Australia (2011), Brazil (2011), Germany (2012), and Spain (2013). Under each of these agreements, FAA agreed to share information about R&D efforts, fuel testing or approval requirements, and environmental or sustainability studies, among other things. According to FAA, these international partnerships contribute to FAA's ongoing efforts to support approval of additional sustainable alternative jet fuels by ASTM International. For example, representatives from all four countries participated and shared information about their respective initiatives at a recent meeting sponsored by CAAFI. Also, FAA and German Ministry of Transport officials recently participated in a technical and coordination exchange to share details on fuel testing and approval activities to identify complementary activities, among other things.

Price-Competitiveness Is the Main Challenge; Federal Activities Help Address Challenge, but Market Factors Affect the Long-term Commercial Viability of Alternative Jet Fuels

While Some Progress Has Been Made, No Alternative Jet Fuels Are Currently Commercially Available at a Price That Is Competitive with Conventional Jet Fuels

The dates FAA and DOD have established for meeting their alternative jet-fuel usage goals are several years or more away, and, to date, all alternative jet fuels purchased in the United States have been for fuel testing, approval, or demonstration activities, not for day-to-day operations. For example, in November 2011, two domestic airlines purchased alternative jet fuels for a limited number of commercial flights. According to DOD officials, DOD purchased about 150,000 gallons in fiscal year 2012 (about 1.5-million total gallons since fiscal year 2007) of alternative jet fuels, all for fuel testing and approval activities, including about 100,000 gallons for the Navy's Great Green Fleet demonstration in July 2012. Commercial and military use is constrained because alternative jet fuels are not yet produced on a commercial-scale at a price that is competitive with conventional jet fuel. While FAA officials acknowledged that FAA's usage goal is "aspirational," they noted that alternative jet fuel use could increase substantially once the industry is capable of producing alternative jet fuels at a commercial scale and at a price that is competitive with conventional jet fuels.

Currently, the price for alternative jet fuels exceeds that of conventional jet fuel. Jet fuel end users—both commercial airlines and DOD—are extremely price sensitive when making purchasing decisions. Fuel purchasers are either unwilling to pay a premium for alternative jet fuels as compared to conventional jet fuel, or in the case of DOD, are precluded by law and department policy from doing so.[43] The actual price differential depends on the feedstock, the production process used to produce the alternative jet fuel, as well as fuel distribution and quantities produced. Of the two alternative jet-fuel production processes approved for use in commercial and military aircraft (Fischer-Tropsch and HEFA), DOD, according to a DOD official, paid from about $3 to $150 per gallon.[44] These prices, however, reflect purchases of small quantities of fuel for testing and approval activities, which according to government officials and a fuel producer we interviewed and literature we reviewed, are higher than what the price would be if the quantities were produced at a commercial scale. A study conducted by one of FAA's COEs in March 2013 estimated that alternative jet fuels produced on a commercial scale using the HEFA process would require a subsidy of $0.35 to $2.86 per gallon to be price-competitive with conventional jet fuels in 2020.[45]

Recent developments indicate that alternative jet fuel use may increase in the future, which could contribute to achieving FAA's and DOD's usage goals. For example, as of January 2014, seven new potential alternative jet-fuel production processes are undergoing review by ASTM International for approval—at least one of which FAA officials told us may be approved by June 2014. According to a couple of government officials and an industry representative whom we spoke with, a range of approved production processes could diversify and expand the future supply of alternative jet fuels. Another potential alternative jet-fuel production process that will be submitted to ASTM for approval involves using a type of renewable fuel that is currently used in ground transportation. Because production capacity already exists for this fuel, it could be made available more quickly to meet demand from the aviation industry. In addition, two airlines—United Airlines and Alaska Airlines—have entered into agreements, known as "off-take agreements," to purchase alternative jet fuels from fuel producers' future production.[46]

The Overarching Challenge to Developing an Alternative Jet Fuels Market Is Making the Price Competitive with Conventional Jet Fuels

We interviewed 23 academic, federal government, and private industry stakeholders with expertise in various segments of the supply chain to help identify challenges to developing and using alternative jet fuels (see app. I for more information on the criteria used to select the stakeholders interviewed).

Through interviews with these stakeholders using open-ended questions, we identified two major factors—high development costs for alternative jet fuels and uncertainty with respect to federal regulations and policies—as the primary contributors to the overarching challenge that alternative jet fuels are not commercially available at a price that is competitive with conventional jet fuel.

High Development Costs across the Alternative Jet Fuel Supply Chain

Almost all of the stakeholders whom we interviewed (22 of 23) cited at least one factor related to high development costs.

While one of these stakeholders discussed the challenges associated with high development costs broadly, the remaining 21 of them highlighted development cost challenges associated with specific supply-chain segments.

- **Feedstock production:** Stakeholders we interviewed most commonly cited the high cost of feedstock in connection with the first segment of the supply chain (15 of 23 stakeholders). Five of these stakeholders noted that for fuel produced using the HEFA production process, the cost of some types of feedstock—even before it is transported or converted—currently exceeds that of conventional fuel. For example, when comparing conventional jet-fuel prices reported by the Energy Information Administration and soybean oil prices reported by the World Bank between 1990 and 2012, the price per gallon of soybean oil exceeded the price per gallon of conventional jet fuel in almost every year.[47]

  In addition, an increase in demand for alternative jet fuels could increase the derived demand for feedstocks (as alternative jet fuel producers increase their production output) and the price of feedstocks could rise.

  Six stakeholders noted that the expansion of low cost natural gas production in the United States could help lower production costs for alternative fuels derived from nonrenewable sources, such as natural gas. However, jet fuel produced from nonrenewable sources, such as natural gas, does not meet the statutory definition of "renewable biomass,"[48] and therefore could not generate RINs. Expanding the production of different types of renewable feedstocks could also help lower feedstock production costs, but three stakeholders noted that the agriculture community does not have much experience in growing crops that could be used to produce alternative fuel, and farmers are hesitant to grow those crops without a guarantee that they can be sold or the certainty that the energy crop would be more profitable than what the farmer could otherwise grow.

- **Feedstock Logistics:** While the logistics differ depending on the type of feedstock, one private-industry stakeholder and studies we reviewed explained that feedstock used to produce alternative fuels is generally costly to collect, store, handle, and transport. For example, oil producing feedstocks—such as camelina, soy, or jatropha—require special handling, including proper moisture and temperature conditions for storage and cleaning, drying, and de-hulling before the process for extracting the oil from the plant occurs. And, some other feedstock crops, such as wood residues or switchgrass—which are fibrous, have a low energy density, and have variable moisture content—are costly to collect, store, and transport because of this complexity. Moreover, to be more cost effective, these feedstocks may need to be grown near fuel production facilities; otherwise, they may need to be shipped by bulk freight transportation (such as by rail or pipeline), which increases the transportation cost. That is why a demonstration-scale ethanol biorefinery that we visited that anticipates producing alternative jet fuels acquires its feedstock (woody biomass) from a poplar tree plantation less than 10 miles away (see fig. 2). Operators of the biorefinery told us that the close proximity of the feedstock source to the biorefinery helps reduce its feedstock logistics costs.

Source: GAO; © 2013 Erik Isakson & GreenWood Resources (feedstock production photograph); © GreenWood Resources (feedstock logistics photograph); ZeaChem Inc. (fuel production photograph).

Figure 2. Logistics of Transporting Feedstock from Source to Fuel Production Facility.

- **Fuel production:** More than half of the stakeholders we interviewed (14 of 23), as well as literature we reviewed, indicated that the high costs associated with transitioning to commercial scale production—such as the capital costs required to construct a commercial-scale production plant—is a key contributing factor affecting the cost of producing alternative jet fuels. For example, one study conducted by the National Research Council estimated that the costs to construct a single biorefinery converting biomass into a liquid transportation fuel using different conversion technologies range from $200 million to $606 million.[49] Stakeholders (5 of 23) also noted that the capital investment costs for constructing alternative fuel-production plants would be even higher when producing fuel from nonrenewable feedstock, such as natural gas.

  Ten stakeholders we interviewed highlighted fuel producers' difficulty in obtaining the private investment needed to help construct commercial-scale alternative fuel production plants. According to stakeholders and literature we reviewed, private financiers are hesitant to invest, in part, because of risks associated with the uncertainty about access to a steady supply of feedstock, high feedstock and capital costs, and an unwillingness on the part of fuel end users to pay a premium price for alternative jet fuels. This is generally true of many capital-intensive start-ups, including other renewable energy industries; four stakeholders noted to us, for example, that the ethanol industry would not be as commercially viable today without considerable federal support. One private-industry stakeholder whom we spoke to noted that fuel producers learn and adapt their processes as they gain experience building and operating commercial-scale production plants. In other words, once fuel producers construct a commercial-scale plant and begin operating it, they can work to create efficiencies in the production process to reduce costs in other ways. Another private-industry stakeholder underscored that the amount of time and funding it takes to move from a good idea in the lab to a commercial scale of production is substantial. Another stakeholder highlighted the cyclical nature of the challenge—that is, fuel producers typically require outside investment finance to construct a commercial-scale plant that can create efficiencies sufficient to decrease costs, while a private financier is hesitant to invest funds unless the producer can lower development costs and guarantee that the fuel price will be competitive with the price of conventional fuels.

- **Fuel testing and approval:** Ten stakeholders, as well as literature we reviewed, explained that the time and testing requirements associated with the testing and approval process for alternative jet fuels add additional cost—in part because alternative jet fuels, in contrast to alternative fuels used for other purposes such as surface transportation, require a more rigorous testing and approval process. According to an industry report, the ASTM International's testing and approval process can last as long as 3 years and cost upwards of $30 million.[50] One stakeholder explained that the fuel testing and approval process requires producers to demonstrate that their production processes are "robust and repeatable" and can reliably control product quality. With regard to the time required to reach fuel approval, the commercial and military approval process for HEFA generally took about 3 years. One stakeholder highlighted that given the amount of time required to get approval for alternative jet fuels, producers may opt to produce other products, such as diesel, that they can get to market more quickly.[51]

  In addition, as an alternative fuel progresses through the testing and approval process, the sequence of tests—ranging from laboratory tests on the fuels to potentially full-scale aircraft tests—require an increasing quantity of fuel to conduct. Some stakeholders (5 of 23) elaborated that since most fuel producers are generally companies with limited funds and small-scale operations, it is extremely costly for them to produce fuel in large quantities. For both the Fischer-Tropsch and HEFA approval processes, the federal government has in some cases provided funding—including purchasing fuels for testing and providing the equipment needed to conduct the tests—to help relieve some of the costs for producers. Air Force officials estimated that DOD generated about 80 percent of the testing data for past approvals, while private parties are leading most of the current fuel certifications. Four government and private-industry stakeholders whom we interviewed, as well as other FAA and DOD officials, expressed concerns that recent cuts to the Air Force Alternative Fuels Certification Division overseeing testing and approval will add to the time and cost of getting additional alternative jet fuels approved. Specifically, the Air Force's Alternative Fuels Certification Division was eliminated in fiscal year 2013 and the funding for the other Air Force division involved in fuel testing and certification—the Air Force Research Lab—is also being cut. A senior Air Force Energy official noted that the Air Force plans to request that funding for its fuel testing and approval activities be restored once the budget situation has improved. The official did not know what the consequences of the recent cuts would be, but indicated that because none of these alternative jet fuels would be immediately commercially available, the short-term impact would be minimal.

Uncertainty of Federal Policies and Regulations

As discussed above, federal policies and regulations can support the development and use of alternative jet fuels, but uncertainty regarding the future of this federal support may limit the support that these policies provide to the alternative jet-fuels industry. More than half of the stakeholders we interviewed (13 of 23) indicated that continued uncertainty in federal regulations and policy contribute to the overarching challenge of making the price of alternative jet fuels competitive with the price of conventional jet fuels, a situation that undermines the viability of the alternative jet-fuels industry as compared to conventional jet fuel. Specifically, these stakeholders cited uncertainty about the RFS and federal tax expenditures[52] as a challenge to developing and using alternative jet fuels.

- **RFS:** Government, academic, and private-industry stakeholders (8 of 23) highlighted legal and political challenges to the RFS, which creates uncertainty about requirements in the future. These challenges include multiple lawsuits[53] filed regarding the validity of the program's volumetric requirements, and political opposition for the program from some lawmakers.[54] The uncertainty about the future of the RFS contributes to private financiers' hesitancy to invest in the biofuels industry, including alternative jet fuels. Even though renewable jet fuels produced from specified feedstocks and conversion processes can qualify as a renewable fuel to meet the advanced biofuels requirement of the RFS, EPA officials reported that no RINs for the production of renewable jet fuel have been generated as of January 2014.[55] However, four stakeholders noted that alternative jet-fuel producers could use RIN credits to help offset fuel development costs and to make the fuel's price more competitive with the price of conventional jet fuel. Four stakeholders, as well as a private financier whom we spoke with, noted that, generally, private financiers wholly or partly discount any potential RIN credit value when evaluating a fuel producer's financial prospects and deciding whether to invest because of uncertainty about the future of the program. Discounting the value of a RIN credit in financial models makes the investment in an alternative fuel producer look less profitable and overall less attractive than the same investment would look with a stable and fully valued RIN credit. Thus, while the uncertainty does not increase the actual price of alternative jet fuel, it may hinder investments that could make alternative jet fuels more price-competitive to conventional jet fuels.[56] Two other stakeholders pointed to the EPA's recent proposal to reduce the total advanced biofuels standard in 2014 under the RFS as indicative of this uncertainty.[57] Specifically, one stakeholder highlighted that when EPA uses its statutory authority to reduce the statutory standards, for example, for volume under RFS, it creates uncertainty, which ultimately makes supporting the research, development, and production of alternative fuels—including alternative jet fuels—less attractive to private financiers.[58]

- **Tax expenditures:** Two private-industry stakeholders pointed out that tax credits to incentivize alternative jet fuels and general biofuels investment and production are authorized for short periods of time, such as 1 year, and on occasion, were not renewed. For example, since its enactment in 2004, the biodiesel tax credit expired in 2010, 2011, and again, most recently, on December 31, 2013.[59] This introduces uncertainty and, similar to the RIN credits, generally causes private financiers to minimize or discount the tax expenditures' value when assessing fuel producers' future expenses and profitability. Literature we reviewed highlighted the potential for stable federal tax policy to contribute to the growth of the renewable energy sector.[60] For example, one study noted that investments in new commercial scale wind and solar power production facilities were fostered by production and investments tax credits, respectively. We also found that the long-term ethanol tax credit was important in creating a profitable ethanol industry when the industry had to fund investment in new facilities.[61]

Stakeholders Generally Believe Federal Actions Are Needed to Advance the Development of Alternative Jet Fuels, While Market Factors Also Affect Long-term Commercial Viability

Federal Actions

Stakeholders and studies we reviewed identified a variety of actions that could assist in the development of alternative jet fuels, ranging from continuing the current federal efforts to providing greater regulatory and policy certainty to providing greater financial support. Stakeholders identified a variety of federal actions as being the most critical for the federal government to take.

- **Research and development efforts:** A majority of the government, academic, and private-industry stakeholders that we interviewed (19 of 23) generally agreed that the federal government should continue research and development efforts to advance the development and use of alternative jet fuels. Seven of the government, academic and private-industry stakeholders suggested, however, that the federal government could be more targeted in its feedstock research and development efforts, such as finding scientific breakthroughs to converting algae and cellulosic feedstocks into fuels; developing more feedstock options; and identifying what feedstocks may be particularly well-suited to producing alternative jet fuels. Five private-industry stakeholders noted their support of the federal government's current efforts related to fuel testing and approval activities. For example, FAA and DOD have helped or plan to help fund specific fuel tests for seven new potential alternative jet-fuel processes that, as of January 2014, are undergoing review by ASTM International for approval. According to federal agency officials, they anticipate that at least one of these new processes may be approved by June 2014. Four government and private-industry stakeholders proposed that the federal government use and increase access to its jet engine assets for testing purposes or help streamline the fuel-testing and approval process for commercial use. For example, one of these government stakeholders suggested that the federal government use an outside contractor to manage the fuel-testing and approval process with subcontracts with facilities for fuel testing, so fuel samples would be sent to a centralized location. All entities with a role in the fuel-testing and approval process (laboratories, engine and aircraft manufacturers, and administrative functions) could then coordinate and agree upon a final approval determination within several months. Ultimately, this approach, the stakeholder noted, could help shorten the ASTM International's approval timeline, lower the fuel testing costs, and add clarity and regularity to the process. Three stakeholders believed that expanding the range of approved production processes would diversify and expand future supply of alternative jet fuels.

- **Regulatory and policy certainty:** More than half of the government, academic, and private-industry stakeholders we interviewed (15 of 23) generally agreed that the federal government should provide greater regulatory and policy certainty to support the development and use of alternative jet fuels. The most commonly cited actions (12 of 23 stakeholders) were providing greater certainty that the renewable fuel program will not be repealed and changes minimized—specifically, reductions—in mandated volumes from year to year. Multiple stakeholders (8 of 23) told us that this may encourage private financiers to value RIN credits when making investment decisions. Ultimately, these stakeholders believe that this certainty would attract more private investment in the industry and help it advance. Other actions mentioned by stakeholders included expanding the RFS by expanding the types of feedstocks that would generate a RIN or mandating a minimum volume standard for alternative jet fuels. EPA officials told us, however, that because the definition of transportation fuel in Section 211(o) of the Clean Air Act,[62] which pertains to the RFS, does not include jet fuels, having the fuel refiner or importer include the jet fuel they produce or import in determining their annual renewable fuel obligation would require Congress to revise the program. Two stakeholders noted that some tax expenditures that are authorized for short periods of time, such as 1 year, should be authorized for a longer period of time—such as a minimum of 10 years. They said that this would be helpful in spurring additional private investment in producing alternative fuels, including alternative jet fuels. Researchers currently studying the impacts of government subsidies on this industry told us that government mandates, such as the RFS, and tax subsidies, as well as their duration, are key factors in making the production of alternative jet fuels profitable.
- **Direct financial support:** A majority of stakeholders (15 of 23) and literature we reviewed also highlighted that greater financial support for alternative jet fuels derived from both renewable and

nonrenewable sources would advance the industry. The two most common potential federal government actions cited by stakeholders included entering into long-term contracts for fuel purchases (6 government, academic, and private-industry stakeholders) and stabilizing existing federal direct support programs (6 government and private-industry stakeholders). Both of these actions would provide greater certainty and reduce some of the risks of private investment. Currently, DOD has statutory authority to enter into certain contracts procuring services or property, including contracts for the purchase of alternative fuels, for up to 5 years with an option to extend the contracts up to 10 years.[63] However, stakeholders we interviewed, as well as industry experts said that the length of the initial contract period is too short to stimulate the private capital market or to encourage potential alternative fuels suppliers to construct or expand production facilities. Alternative fuels producers have told DOD that initial contracts for fuel purchases of at least 10 years in duration would help advance the industry beyond the small production volumes currently planned. Because this would require a statutory change to DOD's authority to contract for jet fuel purchases, DOD has drafted legislative proposals over the past several fiscal years that would allow it to enter into longer-term contracts. One proposal advanced for congressional consideration, but was not adopted. A senior DOD official told us that under current law, DOD would be required to obligate sufficient funds in the first year of a long-term contract—for example, 10 years in length—to pay for the total guaranteed minimum purchases over the duration of the contract, unless it received a specific statutory exemption to do otherwise. Without such an exemption, according to this official, long-term contracts would have a major effect on DOD's budgets and obligational authority. Also, according to a senior DOD official, obtaining and exercising the authority to enter into longer-term contracts could commit DOD to a particular type of alternative jet-fuel production process, while the technological advancements in this industry are changing quickly and could provide a newer and potentially less expensive production process, which DOD may not be able to take advantage of. This DOD official also noted that conventional fuel providers prefer 1-year fuel purchase contracts. Thus, if DOD had and exercised authority to issue longer-term solicitations for bulk fuel purchase, the department could find itself subjected to a bifurcated procurement strategy where the majority of its fuel contracts would be 1-year contracts with conventional fuel providers and the remainder would be longer-term contracts with alternative fuel providers.

Some stakeholders (6 of 23) suggested ensuring stability in the funding stream for the existing Advanced Drop-In Biofuels Production Project, discussed previously, which was to be jointly funded by the USDA, DOE, and DOD. To date, the funding put toward the project is less than the intended amount of $170 million from each federal agency for an aggregate total of $510 million. According to DOD officials, $100 million in fiscal year 2012 funds were applied to this project. For fiscal year 2013, the explanatory statement for the Consolidated and Further Continuing Appropriations Act, 2013,[64] listed an additional $60 million for this purpose, which was conditioned by a provision in the authorization act providing that the funds appropriated would not be obligated or expended until matching funds were received from DOE and USDA.[65] And, only in fiscal year 2014, did DOE receive specific authorization to contribute $45 million, which will be applied to Phase 2 of the project. According to a senior USDA official, while USDA was apportioned[66] about $23 million in fiscal year 2013 for these activities, it has not expended any funds to date toward this initiative.[67] In May and June 2013, four private fuel producers were selected to receive awards totaling $20.5 million for Phase 1.

The 23 stakeholders whom we interviewed had varying views about the future of the alternative jet-fuels industry if current federal activities continue at the same level and no additional federal government action is taken. Specifically, 5 stakeholders believed that the industry would remain commercially unviable—that is, continue to produce small quantities of fuel at prices that are not competitive to conventional jet fuels' prices. Another 6 stakeholders believed that the alternative jet-fuels industry would continue to progress, but at a slow pace. The remaining 12 stakeholders did not articulate a specific prediction for the future of the alternative jet-fuels industry. Some highlighted their concerns that existing policies and programs—such as, the RFS or the Advanced Drop-In Biofuels Production Project—would be repealed, while others expressed pessimism about the future of the industry if the federal government does not address larger industry-related economic or policy challenges.

Market Factors

More than half of the stakeholders (15 of 23) highlighted that market factors, such as the favorable economics for developing competing products (e.g., diesel), will ultimately be a key factor in determining the long-term success of the alternative jet-fuels industry. These stakeholders highlighted three key market factors—favorable economics for competing end-use products or co-products, dependence on commodity markets, and the cost of conventional jet fuel—that they believe will affect the future prospects for the alternative jet-fuels industry. The remaining 8 stakeholders did not offer comments on market factors.

- **Favorable Economics for Competing End Products:** Nine stakeholders, as well as literature we reviewed, highlighted that, currently, end-use products or co-products (such as diesel fuel, naphtha, cosmetics, and plastics) from the same production processes used to produce alternative jet fuels are often cheaper and easier to produce and therefore more profitable as compared to alternative jet fuels. For example, a study of the HEFA process, funded, in part, through one of FAA's COEs, found that if the production goal was to maximize the total amount of all types of liquid fuel, rather than specifically jet fuel, then less than 13 percent of the product mix output would be jet fuel, while almost 70 percent would be diesel; the remaining product mix would consist of propane, naphtha, and liquefied petroleum gas.[68] Maximizing the amount of jet fuel produced would reduce a fuel producer's profitability due to higher operating costs and lower revenues. Two stakeholders that we spoke with who have expertise in fuel production told us that they choose to produce more of the other end products over alternative jet fuels because the other products are more profitable. For example, one stakeholder stated that he has experience selling renewable diesel in one state at a price premium and market demand is higher for renewable diesel.
- **Dependence on Commodity Markets:** Because some alternative jet fuels are made from tradable commodities, the cost of jet fuel production depends on prices in commodity markets. As noted earlier, the price of soybean oil—an input to alternative fuels—has historically exceeded the price of conventional jet fuel. Consequently, it has been impossible for a producer of alternative jet fuels that uses the HEFA production process and soybean oil as a feedstock to

compete on price alone with conventional jet fuels, even if the producer's other production and transportation costs were negligible. Furthermore, in many instances, the input commodities (feedstock) have alternative uses. For example, oil-producing and cellulosic feedstocks can be used to generate heat, power, and other ground transportation fuels. Therefore, an increase in demand for these feedstocks in alternative uses could raise their price and the costs of producing alternative jet fuels.

- **Cost of Conventional Jet Fuels:** Increases in the supply of conventional jet fuels would make it harder for alternative fuels to compete based on price alone. And although international petroleum markets heavily influence the prices of conventional fuels, which can be volatile and difficult to predict into the future, domestic policies can affect jet fuel supply and fuel prices. For example, five stakeholders highlighted that there has been significant previous federal investment in establishing the conventional petroleum industry, such as through long-standing federal tax expenditures that encourage exploration and drilling for conventional petroleum oil. In addition, two stakeholders highlighted that the price of conventional jet fuel does not reflect its full life-cycle cost. Costs not reflected in the price of conventional jet fuel could include environmental externalities, such as the impact of greenhouse gas emissions from aviation on climate change, or direct negative effects on human health from the combustion of jet fuel. If comparing the full life-cycle costs of conventional versus alternative jet fuels, alternative jet fuel could be more cost competitive. However, there is no globally recognized approach for determining the greenhouse gas effects of renewable fuels and the magnitude of any greenhouse gas reductions attributable to their production and use.

AGENCY COMMENTS

We provided DOT, USDA, DOE, DOD, and EPA with a draft of this report for their review and comment. DOT provided technical comments that we incorporated as appropriate. In addition, in comments emailed to us, DOT highlighted that alongside its federal agency partners, it is fully committed to the development and use of sustainable alternative jet fuels to address the nation's energy security, economic development and environmental needs.

DOT also stated that it believes that it has taken a comprehensive approach to overcome barriers to the development and deployment of sustainable alternative jet fuels that are drop-in replacements to fuels derived from petroleum and that these fuels hold great promise and are an essential component of ensuring that the flying services the nation relies upon today remain affordable and available into the future. USDA, DOE, DOD, and EPA also provided technical comments that we incorporated as appropriate. In addition, EPA provided written comments, stating that the report's findings related to approved jet fuel pathways under RFS are accurate.

Gerald L. Dillingham, Ph.D.
Director
Physical Infrastructure Issues

Zina D. Merritt
Director
Defense Capabilities and Management

APPENDIX I: OBJECTIVES, SCOPE, AND METHODOLOGY

This report examines (1) the role of the federal government in the development and use of alternative jet fuels and (2) key challenges to developing and using alternative jet fuels and actions that the federal government plans to or could take to help address those challenges.

To examine the role of the federal government, we first identified the federal agencies most involved in the development or use of alternative jet fuels through review of relevant documents and preliminary interviews with stakeholders. We selected five federal agencies: the Department of Transportation's (DOT) Federal Aviation Administration (FAA), the Department of Agriculture (USDA), the Department of Energy (DOE), the Department of Defense (DOD) and its military departments (Army, Navy, and Air Force), and the Environmental Protection Agency (EPA). For each of these five selected federal agencies, we interviewed officials and reviewed strategic plans, performance reports, and other relevant documents to obtain information about key federal programs, initiatives, or goals that targeted the development or use of alternative jet fuels. We also reviewed literature related to alternative jet fuels and interviewed other stakeholders, including representatives from the private industry, such as fuel producers and airlines, as well as representatives from public-private partnerships. To identify activities that federal agencies are involved in to coordinate their alternative jet fuel related activities with other federal agencies, private industry, and stakeholders, we reviewed relevant memoranda of understanding; international cooperative agreements; reports from public-private partnerships—such as the Commercial Aviation Alternative Fuels Initiative—and regional initiatives; as well as interviewed officials from the five selected federal agencies and relevant interagency working groups. In addition, to identify broad federal strategies and initiatives related to alternative fuels generally, we reviewed key White House and other relevant government-wide documents, including the National Plan for Aeronautics Research and Development and Related Infrastructure (December 2007); Growing -America's Fuels strategy (February 2010); the Blueprint for a Secure Energy Future (March 2011); the National Bioeconomy Blueprint (April 2012); and the President's Climate Action Plan (June 2013). Finally, we also reviewed applicable federal laws and regulations related to alternative transportation fuels, including the Defense Production Act of 1950;[1] Commodity Credit Corporation Charter Act;[2] Energy Policy Act of 2005;[3] Energy Independence and Security Act of 2007;[4] the Food, Conservation, and Energy Act of 2008;[5] the FAA Modernization and Reform Act of 2012;[6] and EPA's regulations for the renewable fuel program.[7] We focused our study on federal efforts, except to the extent that international or private-sector efforts are coordinated with federal efforts. We did not provide an exhaustive list of federal initiatives; rather, we discussed key programs or initiatives that federal agency officials told us and our review of agency documents identified as playing a key role in supporting the development or use of alternative jet fuels or in achieving related goals.

To examine key challenges to developing and using alternative jet fuels and actions that the federal government plans to or could take to help address those challenges, we (1) identified the extent to which alternative jet fuel has been purchased for commercial and military use in the United States; (2) selected and interviewed 23 stakeholders representing government, academia, and the private sector to obtain their views on key challenges and planned or possible federal actions;[8] (3) reviewed relevant literature on challenges to developing and using alternative jet fuels to help corroborate the views obtained from the 23 stakeholders; and (4) interviewed officials from the five selected federal agencies, as well as representatives from other non-federal entities involved in the alternative jet-fuels industry. To identify the extent to which alternative jet fuel has been purchased for commercial and military use, we obtained alternative jet-fuel purchase information for fiscal year 2012 from

Airlines For America—a domestic airline industry group—and for fiscal years 2007 to 2013 from DOD. While we obtained actual quantities from Airlines For America and from DOD, we did not assess the data's reliability because it is not material to our findings and reported the fuel purchase quantities on an order of magnitude in this report. As part of our methodology for selecting the 23 stakeholders, we first identified a list of potential stakeholders by reviewing background information, including federal agency documents; articles published in scholarly journals; and documents produced from conferences and by regional initiatives related to alternative jet fuels. In addition, we considered names of stakeholders recommended during initial interviews we conducted with federal government officials, as well as representatives from professional associations and private industry. We selected the 23 stakeholders representing government, academia, and private industry based on criteria that included:

- type and depth of experience and knowledge in the area of alternative jet fuels;
- recognition in the professional community;
- relevance of published work to the scope of our review;
- representation of a range of expertise across the alternative jet fuel supply chain, such as feedstock development or fuel production;
- representation of a range of stakeholders with knowledge about alternative fuels derived from renewable, nonrenewable, or both sources; and
- representation of a range of stakeholders with knowledge about alternative fuel use in commercial, military, or both settings.

We asked each of the 23 stakeholders a series of semi-structured, open-ended questions about economic, policy, technological, and other key challenges related to developing and using alternative jet fuels and actions that the federal government plans to or could take to help address them. We synthesized the stakeholders' views to identify categories of key challenges and planned or potential federal actions to help address those challenges. The views of these stakeholders are not generalizeable to those of all stakeholders with expertise in the area of alternative jet fuels; however, we believe that they represent a balanced and informed perspective on the topics discussed. In addition, we reviewed relevant literature obtained through background research and from federal agency officials that discussed challenges to developing and using alternative fuels and planned or potential federal actions to help address those challenges. We also conducted a literature search and reviewed five documents that we identified as relevant to challenges related to developing and using alternative jet fuels. Our literature search targeted bibliographic databases containing content on commercial and defense aviation, energy, or both, including Transportation Research International Documentation (TRID); SciSearch; and the Defense Technical Information Center (DTIC). Within these resources, the search focused on scholarly journal articles, conference papers, government reports, and industry trade press published in 2011 and forward. Through the literature search and review of abstracts, we initially identified 36 documents that potentially discussed economic, policy, technological, or other challenges to developing and using alternative jet fuels. We could not obtain 2 of the 36 documents because they were not readily available. Ultimately, after a review of the remaining 34 documents, we identified 5 relevant to our review and reviewed them to help corroborate the key challenges identified by the 23 stakeholders we had interviewed. For those studies we cited in the report, we reviewed their methods, assumptions, and limitations to ensure that they were sufficiently methodologically sound and determined that they were sufficiently reliable for the purposes of our report. Lastly, we interviewed officials from the five selected federal agencies and representatives from other non-federal entities involved in the alternative jet-fuels industry, including fuel producers, airlines, airframe and engine manufacturers, environmental groups, and a private financier to also obtain their views on key challenges to developing and using alternative jet fuels and actions that the federal government plans to or could take to help address those challenges. We did not rank the planned or potential federal actions to help address the key challenges that were identified.

We conducted this performance audit from February 2013 to May 2014 in accordance with generally accepted government auditing standards. Those standards require that we plan and perform the audit to obtain sufficient, appropriate evidence to provide a reasonable basis for our findings and conclusions based on our audit objectives. We believe that the evidence obtained provides a reasonable basis for our findings and conclusions based on our audit objectives.

End Notes

1 The Intergovernmental Panel on Climate Change is a United Nations organization that assesses scientific, technical, and economic information on climate change.

[2] According to a 2009 Environmental Protection Agency study, carbon dioxide emissions from domestic aviation accounted for about 3 percent of the total U.S. carbon dioxide emissions in 2009.

[3] GAO, *Aviation and Climate Change: Aircraft Emissions Expected to Grow, but Technological and Operational Improvements and Government Policies Can Help Control Emissions*, GAO-09-554 (Washington, D.C.: June 8, 2009) and *Aviation and the Environment: NextGen and Research and Development Are Keys to Reducing Emissions and Their Impact on Health and Climate.* GAO-08-706T (Washington, D.C.: May 6, 2008).

[4] There is no globally agreed-upon approach, however, for determining the greenhouse gas effects of renewable fuels and the magnitude of any greenhouse gas reductions attributable to their production and use.

[5] Specifically, the retail price of jet fuel per gallon in 2002 was $0.92 in 2014 dollars, and the price of jet fuel per gallon in 2012 was $3.19 in 2014 dollars.

[6] In addition to encouraging the development of an alternative fuels industry, the federal government continues to support a variety of other energy sectors.

[7] Pub. L. No. 109-58, § 1501, 119 Stat. 594, 1067-1076 (2005), codified as amended at 42 U.S.C. § 7545(o) (2014).

[8] EPA has adopted regulations establishing a Renewable Fuel Standard (RFS). 40 C.F.R. ch. 1, subpart K. EPA annually issues regulations that establish the percentage of transportation fuels that fuel refiners, importers, and other obligated parties must ensure are renewable fuels. The law generally requires that surface transportation fuels contain 9- billion gallons of renewable fuels in 2008, and that volumes increase to 36-billion gallons by 2022. *See* 42 U.S.C. § 7545(o).

[9] Advanced biofuels are defined as renewable fuel other than ethanol derived from corn starch that meet certain criteria, including life-cycle greenhouse gas emissions, as determined by the EPA, that are at least 50 percent less than baseline life-cycle greenhouse gas emissions. Lifecycle emissions include all stages of fuel and feedstock production and distribution, from feedstock generation or extraction through the distribution and delivery to and use of the finished fuel by the ultimate consumer.

[10] In 2010, the International Civil Aviation Organization—an advisory organization affiliated with the United Nations that aims to promote the establishment of international civil aviation standards and recommended practices and procedures—set a goal for the global aviation industry to achieve carbon neutral growth from 2020 onward. Two airline industry groups—the International Air Transportation Association and Airlines for America—also established a goal of achieving carbon neutral growth by 2020 onward. Also, the Department of Transportation's Federal Aviation Administration adopted a goal of returning carbon emissions to their 2005 levels by 2020.

[11] Energy Independence and Security Act of 2007, Pub. L. No. 110-140, § 526,121 Stat. 1492, 1663 (2007), codified at 42 U.S.C. § 17142, states that "no Federal agency shall enter into a contract for procurement of an alternative or synthetic fuel, including a fuel produced from nonconventional petroleum sources, for any mobility-related use, other than for research or testing, unless the contract specifies that the life-cycle greenhouse gas emissions associated with the production and combustion of the fuel supplied under the contract must, on an ongoing basis, be less than or equal to such emissions from the equivalent conventional fuel produced from conventional petroleum sources." This law does not specify how to measure life-cycle greenhouse gas emissions.

[12] ASTM International, formerly known as the American Society for Testing and Materials, is a globally recognized leader in the development and delivery of international voluntary consensus standards.

[13] ASTM Standard D7566 covers the manufacture of jet fuel containing synthesized hydrocarbons (those not derived from petroleum hydrocarbons) for commercial use. Manufacture of synthesized jet fuel for military-grade, Jet Propellant-8, is covered in military standard, MIL-DTL-83133H. Synthesized jet fuel for military grade, Jet Propellant-5 is covered in both ASTM Standard D7566 and military standard, MIL-DTL-5624V.

[14] Under the Fischer-Tropsch process, biomass (such as switchgrass or wood waste), coal, or natural gas is processed through gasification and that gas is then converted to synthetic liquid fuels.

[15] Under a HEFA process (or Hydroprocessed Esters and Fatty Acids process), renewable oil (e.g., vegetable oils, animal fat, waste grease, and algae oil) is processed using hydrogen treatment (hydroprocessing) to yield a fuel in the distillation range of jet fuel and diesel.

[16] In 2009, the President made a commitment to reduce U.S. greenhouse gas emissions in the range of 17 percent below 2005 levels by 2020 if leaders from all other major economies agreed to limit their emissions as well.

[17] Other federal agencies, including the National Aeronautics and Space Administration (NASA) and the Department of the Treasury, also play a role. For example, NASA recently launched the Alternative Fuel Effects on Contrails and Cruise Emissions research, which includes a series of flights using the agency's DC-8 flying laboratory to study the effects of alternative biofuel on engine performance, emissions, and aircraft-generated contrails at altitude. And the Department of the Treasury administers tax expenditures—tax credits or special exclusions, exemptions or deductions through which the government provides an indirect subsidy by foregoing tax receipts. GAO, *A Glossary of Terms Used in the Federal Budget Process,* GAO-05-734SP (Washington, D.C.: 2005) 94. The production of some alternative jet fuels may qualify, for example, for biodiesel credit and the special allowance for second generation biofuel plant property. However, by their nature, tax expenditures represent a substantial federal financial commitment.

[18] 42 U.S.C. § 7545(o)(1)(L). EPA issued its final rule for the expanded RFS in March 2010. *See* 75 Fed. Reg. 14670 (Mar. 26, 2010).

[19] Specifically, according to the Energy Information Administration jet fuel represents about 11 percent of all petroleum consumed by the transportation sector in 2012.

[20] 75 Fed. Reg. 14683 (Mar. 26, 2010).

[21] 78 Fed. Reg. 14190, 14682, 14685-14686 (Mar. 5, 2013).

[22] DOT's strategic plan for fiscal years 2012-2016 includes a goal to advance environmentally sustainable policies and investments that reduce carbon and other harmful emissions from transportation sources. Within that strategic goal, DOT states its support for advancing alternative fuels for aviation. FAA also has other emissions goals, such as the U.S. aviation industry returning its carbon emissions to their 2005 levels by 2020, which are supported by the advancement of some alternative jet fuels. Because achievement of these goals is not solely dependent on alternative jet fuels, we do not consider them to be specific goals for advancing the use of alternative jet fuel.

[23] Although the Army uses jet fuel in its ground vehicles and aircraft and participates in DOD's efforts to test and approve alternative jet fuel for use in this equipment, the Army does not have specific alternative jet-fuel-usage goals in its operational energy policy. In addition, DOD's Office of the Assistant Secretary for Operational Energy Plans and Programs—the office responsible for leading and overseeing operational energy plans and programs which

includes establishing the operational energy strategy and monitoring and reviewing all operational energy initiatives in DOD—has established the high-level goal of expanding DOD's operational energy supply options in the 2011 *Energy for the Warfighter: Operational Energy Strategy*. The strategy also includes the military departments' steps to certify aircraft to use alternative jet fuel as one component to achieve this goal.

[24] Navy, Office of the Chief of Naval Operations, *A Navy Energy Vision for the 21st Century* (Washington, D.C.: October 2010).

[25] The Navy includes two broad categories for its energy consumption: afloat and ashore. The Navy uses almost 75 percent of its total energy afloat and over half of that relies upon liquid petroleum-based fuels.

[26] This quantity includes international alternative-fuels purchases.

[27] U.S. Air Force, *U.S. Air Force Energy Strategic Plan* (March 2013).

[28] USDA and DOE administer a variety of programs that address issues related to feedstock production, feedstock logistics, and fuel production. Officials told us that although some projects supported by these programs had an alternative jet-fuel component, they generally had a broader focus. For this reason we did not highlight them here.

[29] CLEEN is part of the Next Generation Air Transportation System (NextGen)—FAA's program to transform the nation's ground-based air-traffic control system to an air-traffic management system using satellite-based navigation and other advanced technology. In addition to supporting alternative jet-fuel testing, CLEEN also supports environmentally promising aircraft technologies.

[30] PARTNER is a cooperative research organization sponsored by FAA, NASA, DOD, EPA, and Transport Canada that includes 12 collaborating universities and about 50 advisory board members who represent aerospace manufacturers, airlines, airports, state and local governments, and professional and community groups.

[31] Pub. L. No. 112-95, § 911, 126 Stat. 11, 142.

[32] This is the total cost, incurred over the past 7 years and is not adjusted for inflation. According to DOD officials, the funding source of the quantities purchased included each military department's Research, Development, Test, and Evaluation funds and the Department of the Navy's Operations and Maintenance funds, and other funds DOD identified as connected with the American Recovery and Reinvestment Act of 2009, Pub. L. No. 111-5, 123 Stat. 115 (2009).

[33] According to the Senate conferees, money was included in the Department of Defense, Military Construction and Veterans Affairs, and Full-Year Continuing Appropriations Act of 2013 (Pub. L. No. 113-6, 127 Stat. 198 (2013)) for research, development, test and evaluation to reduce emissions of coal-to-liquid fuels to meet the goals established in the DOD's Operational Energy Strategy and its Implementation Plan. *See* 159 Cong. Rec. S1497 (Daily Ed., Mar. 11, 2013) (explanatory statement).

[34] Agencies signed the MOU in response to the President's March 2011 *Blueprint for a Secure Energy Future*, which challenged the Secretaries of Agriculture, Energy and the Navy to investigate how they can work together to speed the development of "drop-in" biofuels substitutes for diesel and jet fuel, citing that competitively priced drop-in biofuels could help meet the fuel needs of the Navy, as well as the commercial aviation and shipping sectors.

[35] Pub. L. No. 81-774, codified at 50 U.S.C. Appx. The Defense Production Act of 1950 was enacted to ensure the availability of industrial resources to meet defense needs. Amendments to the Act allow the funds contributed under its authority to be used for energy supply, emergency preparedness, and critical infrastructure protection and require

agencies to report on foreign offsets, which are incentives to foreign governments to purchase U.S. goods and services.

[36] The Commodity Credit Corporation Charter Act, Act of June 29, 1948, ch. 704, 62 Stat. 1070, codified as amended at 15 U.S.C. ch. 15, established the Commodity Credit Corporation within the USDA for the purpose of stabilizing, supporting, and protecting farm income; assisting in the maintenance of an adequate supply of agricultural commodities; and facilitating the orderly distribution of such commodities. USDA, through the Commodity Credit Corporation, has access to capital stock and is authorized to support agricultural commodity prices through loans, purchases, and payments; procure agricultural commodities to meet domestic requirements; purchase agricultural commodities for sale to federal agencies, foreign governments, and international relief agencies; and carry out conservation or environmental programs authorized by law, among other things.

[37] Phase 1 included awards for planning and preliminary designs for biofuel production facilities. Phase 2—available to those entities that participated in Phase 1—would include awards for construction, commissioning, and performance testing of biofuel production facilities.

[38] While DOD has the authority under appropriate circumstances to enter into certain contracts for procurement of services or property, including contracts for the purchase of alternative fuels for up to 5 years with options to extend the life of such contracts to 10 years (10 U.S.C. § 2304a), officials stated that no such long-term contracts have been awarded to date. Instead, DOD has used 1-year purchasing contracts with options.

[39] According to BETO officials, DOE uses an active project management process to ensure that projects meet key milestones, timelines, and budget parameters. When projects are not able to meet theses parameters, DOE discusses the issue with grantees and gives them a chance to address the issues. If DOE and the grantee cannot agree upon appropriate action to address the issues, they may agree to mutually terminate the project.

[40] CAAFI is sponsored by FAA and three trade associations: Airports Council International-North America, Aerospace Industries Association, and Airlines for America.

[41] A domestic airline industry group.

[42] These initiatives include Sustainable Aviation Fuels Northwest and the Midwest Aviation Sustainable Biofuels Initiative, both of which focus on accelerating the development of commercially viable, renewable alternative jet fuels.

[43] DOD may not obligate or expend funds made available for fiscal year 2014 to make a bulk purchase of alternative jet fuel for operational use unless the price of that fuel is competitive with the price of conventional jet fuel. *See* National Defense Authorization Act for Fiscal Year 2014, Pub. L. No. 113-66 § 315, 127 Stat. 672, 730 (2013), Under agency policy, DOD may purchase alternative jet fuel at prices above conventional jet fuel if the fuels are used for research, development, and testing purposes—not for operational use.

[44] Specifically, a DOD official reported that the department paid a range of prices from about $3 per gallon for 315,000 gallons of alternative jet fuel derived from natural gas produced using the Fischer-Tropsch process to about $150 per gallon for 1,500 gallons of alternative jet fuel derived from algal oil produced using the HEFA process.

[45] Partnership for AiR Transportation Noise and Emissions Reduction An FAA/NASA/Transport Canada sponsored Center of Excellence, Market Cost of Renewable Jet Fuel Adoption in the United States (A PARTNER Project 31 Report) (Cambridge, MA: March 2013). In this study, the estimated subsidy depends on the model's underlying assumptions, such as the quantity of renewable jet fuel produced from rotation crops. Under one set of assumptions, called the "Reference Scenario," no HEFA jet fuel is produced in 2020, and the estimated price of conventional jet fuel is $3.41 per gallon. Four other scenarios resulted in the

production of 1 billion gallons of HEFA jet fuel in 2020; in three of these scenarios the implicit per-gallon subsidy to renewable jet fuels exceeds 60 percent of the average price of jet fuel per gallon.

[46] United Airlines and AltAir Fuels announced an off-take agreement in June 2013 that includes the airline's agreement to purchase 15 million gallons of renewable alternative jet fuels over 3 years beginning in 2014. Alaska Airlines and Hawai'i BioEnergy announced an agreement in July 2013 to purchase an unknown quantity of renewable alternative jet fuels as early as 2018.

[47] The average difference between a gallon of soybean oil and a gallon of conventional jet fuel between 1990 and 2012 was $1.19 in 2010 dollars. *See* Partnership for AiR Transportation Noise and Emissions Reduction An FAA/NASA/Transport Canada sponsored Center of Excellence, Market Cost of Renewable Jet Fuel Adoption in the United States (A PARTNER Project 31 Report) (Cambridge, MA: March 2013).

[48] 42 U.S.C. § 7545 (o)(1)(l).

[49] National Research Council of the National Academies, *Renewable Fuel Standard: Potential Economic and Environmental Effects of U.S. Biofuel Policy* (Washington, D.C.: 2011).

[50] Midwest Aviation Sustainable Biofuels Initiative, *Fueling a Sustainable Future for Aviation* (2013). This report did not provide a range of the time and cost of the ASTM International certification and approval fuel process, only an upward estimate.

[51] As discussed above, other end products or co-products—such as, diesel fuel, cosmetics, and plastics—are produced through the same production process.

[52] Tax expenditures are provisions, such as tax credits or special exclusions, exemptions or deductions that reduce taxpayers' tax liability as a loss of federal revenue. See footnote 17.

[53] For example, see *American Petroleum Inst. v. EPA*, 706 F.3d 474 (D.C. Cir., 2013); *Monroe Energy, LLC v. U.S. Environmental Protection Agency*, No. 14-1014 (D.C. Cir., *filed* Jan. 28, 2014).

[54] For example, see H.R. 1461, *Renewable Fuel Standard Elimination Act* (introduced Apr. 10, 2013); S. 1195, *Renewable Fuel Standard Repeal Act* (introduced June 20, 2013); and H.R. 426, *Remove Incentives for Producing Ethanol Act of 2011* (introduced Jan. 25, 2011).

[55] As discussed earlier in this report, fuel producers or importers can obtain RINs by purchasing and blending renewable fuels themselves, or they can purchase RINs from renewable fuel producers, importers, blenders, or other parties. In this way, the RFS has created a market for RIN credits. While jet fuel is not used to calculate a fuel producer's or importer's obligation with respect to renewable fuel, alternative jet fuels produced through approved processes can generate RINs, which the fuel producer may then sell to others, and their sale may help subsidize the cost of producing qualifying alternative jet fuels.

[56] The average RIN credit value for biomass-based diesel—which we use as a proxy for renewable jet fuels since no RINs for the production of renewable jet fuel have been generated—between September 2013 and February 2014 was $0.40 per RIN. RIN credit values ranged from $0.26 to $0.61 per RIN during this time period.

[57] In proposing the 2014 volume standards, EPA for the first time exercised its authority to reduce the statutory standards for advanced biofuels, which has raised additional concern among the renewable fuels industry.

[58] EPA annually issues regulations that establish the percentage of conventional transportation fuels that refiners, importers, and other obligated parties must ensure are renewable fuels. The Administrator of the EPA is authorized to waive the RFS levels established under EISA under certain circumstances. *See* 42 U.S.C. § 7545(o)(7).

[59] Subsequent to the expiration of the credit in 2010 and 2011, when Congress later extended the credit, Congress also made it retroactive back to the date of expiration.

[60] National Renewable Energy Laboratory, *§1603 Treasury Grant Expiration: Industry Insight on Financing and Market Implications*. (Golden, CO: June 2012).

[61] GAO, *Opportunities to Reduce Potential Duplication in Government Programs, Save Tax Dollars, and Enhance Revenue*, GAO-11-318SP (Washington, D.C.: March 1, 2011)

[62] 42 U.S.C. § 7545(o).

[63] *See* 10 U.S.C. § 2304a.

[64] 159 Cong. Rec. S1456 (daily ed. Mar. 11, 2013); see, also, Pub. L. No. 113-6, § 4, 127 Stat. 198, 199 (2013).

[65] National Defense Authorization Act for Fiscal Year 2013, Pub. L. No. 112-239, § 315, 126 Stat. 1632, 3693 (2013) (requiring matching funds before money was obligated or expended for biofuels refinery construction).

[66] Apportionment is the action by which the Office of Management and Budget distributes amounts available for obligation to executive branch agencies, including budgetary reserves established pursuant to law, in an appropriation or fund account. An apportionment divides amounts available for obligation by specific time periods (usually quarters), activities, projects, objects, or a combination thereof. The amounts so apportioned limit the amount of obligations that may be incurred.

[67] According to a senior USDA official, the federal sequestration reduced USDA's initial intended contribution of $170 million to about $161 million. In this regard, the Balanced Budget and Emergency Deficit Control Act (BBEDCA) established discretionary spending limits for fiscal years 2012 to 2021. 2 U.S.C. § 901, as amended. BBEDCA required the Office of Management and Budget to calculate, and the President to order, a sequestration, i.e., cancellation, of budgetary resources on March 1, 2013. *See* 2 U.S.C. § 901a. For fiscal year 2013, the President ordered a sequestration of about $85 billion. *Sequestration Order for Fiscal Year 2013 Pursuant to Section 251A of the Balanced Budget and Emergency Deficit Control Act, As Amended*, 78 Fed. Reg. 14633 (Mar. 6, 2013).

[68] M. Pearlson, C. Wollersheim, and J. Hileman, "A Techno-Economic Review of Hydroprocessed Renewable Esters and Fatty Acids for Jet Fuel Production," *Biofuels, Bioprod, Bioref*, vol. 7:89-96 (2013).

End Notes for Appendix I

[1] Pub. L. No. 81-774, codified at 50 U.S.C. Appx.

[2] Act of June 29, 1948, ch. 704, 62 Stat. 1070, codified as amended at 15 U.S.C. ch. 15.

[3] Pub. L. No. 109-58, § 1501, 119 Stat. 594, 1067-1076 (2005), codified as amended at 42 U.S.C. § 7545(o) (2014).

[4] Pub. L. No. 110-140, § 526,121 Stat. 1492, 1663 (2007), codified at 42 U.S.C. § 17142.

[5] Pub. L. No. 110–234, May 22, 2008, 122 Stat. 923.

[6] Pub. L. No. 112-95, 126 Stat. 11.

[7] 40 C.F.R. ch. 1, subpart K.

[8] Government stakeholders included employees of federal government agencies and a municipal airport and seaport authority. In addition, one expert we interviewed had a colleague present during the interview that shared all of the same views, so for the purpose of this report, we treated them as one stakeholder.

In: Alternative Jet Fuels
Editor: Jenna L. Aiken
ISBN: 978-1-63321-626-6
© 2014 Nova Science Publishers, Inc.

Chapter 2

THE FEASIBILITY OF PRODUCING AND USING BIOMASS-BASED DIESEL AND JET FUEL IN THE UNITED STATES*

A. Milbrandt, C. Kinchin and R. McCormick

ABSTRACT

This study summarizes the best available public data on the production, capacity, cost, market demand, and feedstock availability for the production of biomass-based diesel and jet fuel. It includes an overview of the current conversion processes and current state-of-development for the production of biomass-based jet and diesel fuel, as well as the key companies pursuing this effort. The discussion analyzes all this information in the context of meeting the Renewable Fuels Standard mandate, highlights uncertainties for the future industry development, and key business opportunities.

ABBREVIATIONS, ACRONYMS, AND DEFINITIONS

ATJ	alcohol-to-jet fuel
ASTM	American Society for Testing and Materials

* This is an edited, reformatted and augmented version of Technical Report NREL/TP-6A20-58015, issued by the National Renewable Energy Laboratory, December 2013.

CAFE	Corporate Average Fuel Economy
C&D	construction and demolition
DoD	Department of Defense
EIA	Energy Information Administration
EISA	Energy Independence and Security Act
EPA	Environmental Protection Agency
EtOH	Ethanol
FAME	fatty acid methyl esters
FOG	fats, oils, and greases
FT	Fischer Tropsch
gge	gasoline gallon equivalent
GHG	greenhouse gas
ha	hectare
HEFA	hydrogenated esters and fatty acids
IPK	iso-paraffinic kerosene
LPG	liquefied petroleum gas
mpg	miles per gallon
MSW	Municipal solid waste
PADD	Petroleum Administration for Defense District
R&D	research and development
RFS	Renewable Fuels Standard
TAG	triacylglycerol
TEA	techno-economic analysis

EXECUTIVE SUMMARY

In recent years, growing concern about U.S. dependence on oil imports and greenhouse gas (GHG) emissions has resulted in greater interest in alternative fuels that can be produced from domestic renewable feedstock. Currently, diesel and jet fuel substitutes derived from biomass material—such as crop and forest residues, dedicated energy crops, and plant oils—are receiving increasing attention.

The purpose of this study is to evaluate the technical and economic feasibility of producing and using biomass-based diesel and jet fuel in the United States. To achieve this goal, data on production, capacity, cost, market demand, and feedstock availability were gathered and analyzed. Environmental considerations and legislative climate were not evaluated here, but these factors would contribute to an important follow-up study and provide a more complete picture of the biomass-based diesel and jet fuel potential in the United States.

Some of the key findings of this study include the following:

1) It is technically feasible to produce biomass-derived diesel and jet fuel substitutes in the United States. Many conversion technology options exist. Some are commercially available or in demonstrational stage; others are still in the research and development phase.
2) Biodiesel, consisting of fatty acid methyl esters (FAME) produced from lipids (fats, oils, and greases), is currently the predominant form of biomass-based diesel. Production reached a record 1.1 billion gallons in 2011 and kept at that level in 2012. It is expected to be higher in 2013. Biodiesel blends cannot yet be considered fully "drop-in" fuels because they cannot be transported in all petroleum product pipelines. For pipelines that transport jet fuel, there is a concern that the jet fuel will be contaminated with biodiesel, making it unsuitable for use. Ongoing research aims to determine what, if any, level of FAME can be tolerated in jet fuel.
3) In comparison, the current U.S. renewable diesel and jet fuel production capacity is small, about 225 million gallons per year. These fuels can be produced from various biomass resources and through several different approaches which all target hydrocarbon products that are similar to petroleum fuels in chemical makeup, and therefore may be considered "drop-in" fuels. It is anticipated that, as "drop-in" fuels, they can be blended with petroleum diesel/jet fuel at high levels, or possibly used in neat form.
   - Renewable diesel is produced at commercial scale primarily by hydroisomerization of lipid feedstock. Currently, there are two commercial facilities utilizing this process: Dynamic Fuels, a joint venture between Syntroleum Corporation and Tyson Foods (Geismar, Louisiana) and Diamond Green Diesel, a joint venture between Valero subsidiary Diamond Alternative Energy LLC and Darling International Inc. (Norco, Louisiana).
   - A number of processes are under development for production of renewable diesel from biomass-derived sugars (corn, sugarcane, and sorghum, as well as sugars from thermochemical or biochemical depolymerization of cellulose and hemicellulose).
   - Processes are also being developed for direct conversion of lignocellulosic biomass by fast pyrolysis, gasification, and other

thermochemical means. The first commercial plant is operated by KiOR in Columbus, Mississippi and became operational in early 2013.

4) The costs for producing renewable diesel and jet fuel are not well known and involve a high degree of uncertainty. The process economics for these fuels is highly dependent upon the cost of the feedstock, similar to biodiesel. Additionally, variables such as plant size and co-product credits can have a significant impact on the overall production cost. Hydroisomerization of lipids is performed commercially by Dynamic Fuels and Diamond Green Diesel and internationally by Neste Oil. The KiOR technology, utilizing pyrolysis, is at an initial commercial scale. Other technology routes are not yet commercial and display a wide range of estimated costs in public sources. As these technology pathways mature and become more widespread, more specific information regarding their economics will be available, which will enable a more detailed analysis and performance comparison.
5) From a feedstock perspective, enough lignocellulosic material is projected to be available in support of the Renewable Fuels Standard (RFS) mandate of 21 billion gallons of advanced biofuels. Crop and forest residues alone could yield about 8-24 billion gallons of biomass-based diesel/jet fuel in 2022 (assuming a conversion via fast pyrolysis). This potential could be larger if the conversion technologies achieve higher yields and if additional feedstock, such as dedicated energy crops, become available. However, there will be competition for lignocellulosic feedstock with the ethanol industry and renewable gasoline producers to meet the RFS mandate. Thus, it is unclear what share the renewable diesel/jet fuel would have in the total biofuels contribution. Ultimately, it will depend on the rate of commercialization of these technologies, selling price, and the transportation market demands.
6) Based on current statistics, and proven by the biodiesel industry, there is enough lipid feedstock to support the production of 1 billion gallons of biomass-based diesel mandated by the RFS. Today, roughly half of the biodiesel in the United States is produced from soybeans. The remaining portion consists of animal fat, used cooking oil, canola, and some other minor feedstocks. While soybean production is projected to grow in coming years, the biodiesel industry hopes to achieve higher output through advanced technologies for increasing oil supply

and production of new feedstock. If algal oil becomes commercially available, as projected within the next 5-10 years, it would greatly benefit both biodiesel and renewable diesel/jet fuel industries. Given the right resources, algal oil productivity can be quite high. Algae are a potential aquatic oil crop, but may also yield carbohydrates that can be converted to sugar.

7) Demand for diesel and jet fuel in the United States is projected to grow. As easily recoverable crude oil resources are diminishing and as their prices rise, more substitutes are expected to enter the market.
   - Among the diesel consumers in the country, freight trucking has the largest share. The number of light-duty vehicles using diesel is projected to increase, the rate of which will depend on the market penetration of other alternatives such as hybrid and electric vehicles.
   - Jet fuel is forming as a large and profitable market for the renewable fuels industry. Success in this area could stimulate a significant increase in the production of biofuels and associated feedstock. It is expected that jet fuel consumption by commercial carriers will continue to grow over the next years, whereas jet fuel consumption by the military will remain flat.
8) For biomass-based diesel and jet fuel to be successful among the trucking and aviation companies, they must be cost-competitive with petroleum-based fuels. It is uncertain what the future holds for these substitutes, but it is expected that the next several years, as more facilities come online, will answer many questions about the economic viability of these technologies. Much will depend on the rate of recovery of U.S. and world economies, oil prices, carbon market, and political climate.

INTRODUCTION

In recent years, growing concern about U.S. dependence on oil imports and greenhouse gas (GHG) emissions has resulted in greater interest in alternative fuels that can be produced from domestic renewable feedstock. Currently, diesel and jet fuel substitutes derived from biomass material—such as crop and forest residues, dedicated energy crops, and plant oils—are receiving increasing attention.

The purpose of this study is to evaluate the technical and economic feasibility of producing and using biomass-based diesel and jet fuel in the United States. To achieve this goal, data on production, capacity, cost, market demand, and feedstock availability were gathered and analyzed. Environmental considerations and legislative climate were not evaluated here, but these factors would contribute to an important follow-up task to provide a more complete picture of the biomass-based diesel and jet fuel potential in the United States.

Overview

The Renewable Fuel Standard (RFS), a program introduced by the Energy Independence and Security Act of 2007 (EISA), defines biomass-based diesel as follows: (a) renewable fuel made from biomass; (b) meeting the definition of either biodiesel (mono-alkyl esters) or non-ester renewable diesel; (c) with lifecycle GHG emissions at least 50% less than the diesel fuel it displaces; and (d) which excludes renewable fuel derived from co-processing biomass with a petroleum feedstock[1] (EPA 2009). The jet fuel substitutes fall under the "additional renewable fuel" category of RFS2, defined as fuel produced from renewable biomass that is used to replace or reduce fossil fuels used in home heating oil or jet fuel (EPA 2009).

There are several technologies and processes that produce biomass-based diesel and jet fuel. Some of these technologies are in commercial or pre-commercial production while others are still in the research and development phase. The different technologies use a variety of feedstocks, including lignocellulosic biomass (such as wood and crop residues, dedicated herbaceous or tree energy crops), grains, sugar crops, vegetable oil (soybean, canola/rapeseed, etc.), animal fat (beef tallow, pork lard), and waste cooking greases. Table 1 summarizes the different processes used to create biomass-based diesel and jet fuel, as well as examples of companies involved in these technologies.

Biodiesel is currently the predominant form of biomass-based diesel. It is a liquid fuel made from recycled or virgin vegetable oils and animal fats through a chemical process (transesterification, described below) to produce chemical compounds known as fatty acid methyl esters (FAME). Biodiesel is the name given to these esters when they meet specifications such as ASTM D6751 or European Norm EN14214 for use as transportation fuel. Biodiesel is used in its pure form or in blends with petroleum diesel. Blends containing up to 5% volume are considered the same as conventional diesel and are fully compatible with all engines and infrastructure. Blends containing 6-20% volume biodiesel are accepted by many engine manufactures and are compatible with underground storage tanks. Fuel dispensers and related equipment that are compatible with B6 to B20 blends are available. Biodiesel blends cannot yet be considered fully "drop-in" fuels because they cannot be transported in all petroleum product pipelines. For pipelines that transport jet fuel, there is a concern that the jet fuel will be contaminated with biodiesel, making it unsuitable for use. Ongoing research aims to determine what, if any, level of FAME can be tolerated in jet fuel, and in the meantime, biodiesel is being shipped on pipelines that do not transport jet fuel. A few engine manufacturers approve of the use of B100 (pure biodiesel) in some engine models.

Renewable diesel refers to diesel fuel substitutes derived from biomass sources that chemically are not esters and thus are distinct from biodiesel. Renewable diesel and jet fuel can be produced by several different approaches (Table 1), which all target hydrocarbon products that are similar to petroleum diesel/jet fuel in chemical makeup and therefore may be considered "drop-in" fuels. It is anticipated that, as "drop-in" fuels, they can be blended with petroleum diesel/jet fuel at high levels or possibly be used in neat form. They are also expected to be transportable by pipeline.

Biodiesel from Transesterification

The process of transesterification is used for the conversion of triglycerides (the main component of vegetable oils and animal fats) to biodiesel. In this process, the feedstock is chemically reacted with an alcohol (usually methanol) in the presence of a catalyst like lye. The products are glycerin and the biodiesel fuel or FAME. These separate into a bottom glycerin layer and a top FAME layer, which can then be physically segregated.

Renewable Diesel and Jet Fuel from Hydroprocessing of Lipid Feedstocks

Renewable diesel, sometimes referred to as "green diesel" or hydrogenated esters and fatty acids (HEFA), can be produced from fatty acids (fats, oils, and greases [or FOG]) by the traditional hydroprocessing and hydroisomerization technology used in petroleum refineries. Hydroprocessing is the process of reacting feedstock with hydrogen under elevated temperatures and pressures and in the presence of a catalyst in order to remove oxygen, sulfur, and nitrogen and saturate double bonds.

Table 1. Conversion Processes for Biomass-based Diesel and Jet Fuel

Technology	Process	Fuel Produced	Companies
Lipids (fats, oils, and greases)			
	Transesterification	Biodiesel	Imperium Renewables, Renewable Energy Group, ADM, Amerigreen Energy, Inc., Cargill Inc., Direct Fuels, etc.
	Transesterification of microalgae	Biodiesel	Cellana, Solix, Seambiotic, LiveFuels
	Hydroprocessing	Diesel, Jet Fuel	Neste Oil, Dynamic Fuels LLC, Diamond Green Diesel LLC, UOP, AltAir, Emerald Biofuels LLC, etc.
	Hydroprocessing of microalgae	Diesel, Jet Fuel	Sapphare, Solazyme
Lignocellulosic Biomass/Sugars from Cellulose			
Biochemical	Conversion of cellulose via carboxylic acid	Diesel, Jet Fuel	Terrabon
	Synthetic Biology	Diesel, Jet Fuel	Amyris, LS9, Joule
Thermochemical	Gasification/Fischer Tropsch	Diesel, Jet Fuel	Choren, Flambeau River Biofuels, ClearFuels/ Rentech, TRI, Syntroleum
Hybrid Biochemical/ Thermochemical	Syngas fermentation	Diesel, Jet Fuel	Coskata, Lanztech, INEOS Bio
	Acetic acid production and lignin gasification	Diesel, Jet Fuel	ZeaChem
Depolymerization	Catalytic depolymerization of cellulose	Diesel, Jet Fuel	Covanta, Green Power
	Pyrolysis	Diesel, Jet Fuel	Envergent (UOP/Ensyn), Dynamotive, KIOR, RTI, GTI
	Thermal depolymerization	Diesel, Jet Fuel	Changing World Technologies Inc.
Other	Catalytic reforming of sugars from cellulose	Diesel, Jet Fuel	Virent
	Alchohol-to-jet fuel	Jet Fuel	Gevo, Cobalt

Note: The list of companies in the table is shown as an example and it is not meant to be complete. Some of the listed companies (such as Terrabon and Choren) were inactive at the time of writing. However, these companies are associated with either pioneering or further developing a process, and thus, they are included in this summary for illustrative purposes.

For processing of triglycerides, some of the oxygen can also be removed as carbon dioxide by decarboxylation in some processes (Kalnes et al. 2007). While the feedstock can be hydroprocessed as a co-feed with petroleum, in order to qualify as a renewable fuel (according to the RFS) the diesel must be produced as a dedicated feed in a stand-alone process. The products consist predominantly of isoparaffins with some residual normal paraffins (Smagala et al. 2013). The degree of isomerization can be adjusted to lower the cloud point, and it is even possible to obtain cloud point in the jet fuel range (less than -40°F). However, increased isomerization reduces the cetane number and leads to increased production of naphtha as a by-product. Nevertheless, for fuels with cloud points in the diesel fuel range, the cetane number will likely be over 80.

Renewable Diesel and Jet Fuel from Cellulose via Carboxylic Acid

Terrabon, a company involved in creating renewable fuels using carboxylic acid, describes the process as follows:

> It begins by treating the feedstock with lime to enhance its digestibility, and then fermenting the biomass using a mixed-culture of microorganisms to produce a mixture of carboxylic acids. Calcium carbonate is added to the fermentation to neutralize the acids to form corresponding carboxylate salts, which are then dewatered, concentrated, dried and thermally converted to ketones. The ketones are then hydrogenated to alcohols that can be refined into renewable gasoline, diesel or jet fuel blendstocks (*Ethanol Producer Magazine* 2011, 3).

Renewable Diesel and Jet Fuel Using Synthetic Biology

Synthetic biology modifies existing biological systems or builds new systems to produce novel substances. Amyris (2012) has developed genetic engineering and screening technologies that enable modification of the way microbes process (i.e., metabolize) sugar. By controlling these metabolic pathways, Amyris is able to design microbes, primarily yeast, to be living factories that convert plant-sourced sugars from crops such as sugarcane or sweet sorghum into target molecules. Using its industrial synthetic biology platform, Amyris develops yeast strains designed to produce a broad range of molecules. Amyris's building block molecule is Biofene (an Amyris-brand farnesene), a hydrocarbon molecule that can replace petrochemicals in a wide variety of products including transportation fuels such as diesel and jet fuel.

LS9 has focused on developing renewable petroleum products using a one-step fermentation process. In July 2010, the company announced the discovery of novel genes that, when expressed in *E.coli*, produce alkanes—the primary hydrocarbon components of gasoline, diesel, and jet fuel. This discovery is the first description of the genes responsible for alkane biosynthesis and the first example of a single-step conversion of sugar-to-fuel-grade alkanes by an engineered microorganism. A spokesperson for the company describes the process as a one-step sugar-to-diesel process that does not require elevated temperatures, high pressures, toxic inorganic catalysts, hydrogen, or complex unit operations (Greentech Media 2010). Similarly, the Joule process uses optimized microorganisms that act as living catalysts to produce fuel rather than first producing biomass and later extracting lipids or sugars for a subsequent multistep conversion into fuel (European Biofuels Technology Platform [EBTP] 2012).

Renewable Diesel and Jet Fuel from Biomass Gasification (Fischer-Tropsch)

Another process for making renewable diesel and jet fuel is converting cellulosic biomass through high-temperature gasification into synthetic gas or "syngas", a gaseous mixture rich in hydrogen and carbon monoxide. Next, a Fischer-Tropsch (FT) process is used to catalytically convert the syngas to liquid and wax products that can be refined into synthetic fuels. The production of FT liquids is a commercial technology applied to coal and natural gas.

Renewable Diesel and Jet Fuel from Syngas Fermentation

This hybrid technology is currently used to produce ethanol. However, the companies involved in this process are working on diesel/jet fuel production as well. The strategy involves the gasification of biomass to syngas before processing it into ethanol using a biochemical fermenter. Coskata (2011) describes the process: during gasification, the biomass material is converted into syngas using well-established gasification technologies. After the chemical bonds are broken in the process, Coskata's proprietary microorganisms convert the resulting syngas into ethanol by consuming the carbon monoxide (CO) and hydrogen (H2) in the gas stream. Once the gas-to-liquid conversion process has occurred, the resulting ethanol is recovered from the solution using proven distillation methods.

Renewable Diesel and Jet Fuel from Acetic Acid Production and Lignin Gasification

Similar to the hybrid process above, this technology is currently used to produce ethanol, but the company involved in this process, ZeaChem Inc., has developed a platform capable of producing other fuels as well. After fractionating the biomass, the sugar stream (both xylose [C5] and glucose [C6]) are sent to fermentation where an acetogenic process is utilized to ferment the sugars to acetic acid without CO_2 as a by-product. In comparison, traditional yeast fermentation creates one molecule of CO_2 for every molecule of ethanol. Thus, the carbon efficiency of the ZeaChem fermentation process is nearly 100% vs. 67% for yeast.

The acetic acid is converted to an ester, which can then be reacted with hydrogen to make ethanol. To get the hydrogen necessary to convert the ester to ethanol, ZeaChem takes the lignin residue from the fractionation process and gasifies it to create a hydrogen-rich syngas stream. The hydrogen is separated from the syngas and used for ester hydrogenation, and the remainder of the syngas is burned to create steam and power for the process. The net effect of combining the two processes is that about two-thirds of the energy in the ethanol comes from the sugar stream and one-third comes from the lignin steam in the form of hydrogen (ZeaChem 2011).

Renewable Diesel and Jet Fuel from Catalytic Depolymerization of Cellulose

The catalytic depolymerization process uses heat and catalysts to break long chain polymers of hydrogen, oxygen, and carbon into short-chain petroleum hydrocarbons.

> The Green Power process catalytically depolymerizes cellulosic feedstocks at moderate temperatures into liquid hydrocarbon fuels. The feedstock is first ground to a size finer than 5 mm and then placed--along with a catalyst, a low loading of lime that serves as a neutralizing agent, and a fuel that provides a liquid medium--into a

reactor and heated to around 662°F. In the reactor, the feedstock is catalytically converted to liquid fuels which primarily fall within the gasoline and diesel fuel boiling ranges, although these fuels may need further upgrading. The liquid fuels are separated from any solids which are present and are distilled into typical fuel streams including naphtha, diesel fuel, kerosene, and fuel oil (EPA 2010a, p. 42261).

Renewable Diesel and Jet Fuel from Pyrolysis Oil

Another technique for producing renewable diesel uses pyrolysis, the chemical decomposition of organic materials at elevated temperatures in the absence of oxygen. During this process, large polymers (i.e., cellulose, hemicellulose, lignin, and proteins of organic waste streams) are converted into smaller molecules and produce organic vapors, gases, and a solid residue containing carbon and ash. The vapors are condensed to produce pyrolysis oil (often referred to as bio-oil) that is then refined into diesel-like fuel. Yields of raw bio-oil as high as 75% of the initial dry weight of the biomass can be achieved (PNNL 2009).

Renewable Diesel and Jet Fuel from Thermal Depolymerization

Thermal depolymerization is similar to the geological processes thought to be involved in the production of fossil fuels—except that the technological process occurs in a timeframe measured in hours. Biomass is reacted in water at elevated temperature and pressure to form oils, gases, carbons, and ash. Hydrothermal conversion temperatures are typically 570°-660°F, with pressure sufficient to keep the water primarily as liquid (100-170 atm). The resulting bio-oil could be upgraded to a hydrocarbon product consistent with gasoline and diesel. The technology is being commercialized in the United States by Changing World Technologies (CWT). The National Advanced Biofuels Consortium (NABC) is working on further developing this process. NABC's researchers have achieved bio-oil yields of about 50% on a carbon basis from two feedstocks— wood residues and corn stover (NABC 2012b).

Renewable Diesel and Jet Fuel from Catalytic Reforming of Sugars from Cellulose

Catalytic upgrading of sugars to hydrocarbons involves separating sugars from biomass (e.g., milled corn stover) through a series of chemical and biochemical processes and catalytically upgrading it into hydrocarbon fuels. The process has been researched by Virent, Inc. Virent's BioForming process integrates the company's patented aqueous phase reforming (APR) technology with conventional catalytic processes. First, the lignocellulosic biomass is pretreated, followed by enzymatic hydrolysis (saccharification) of the remaining cellulose. Next is catalytic conversion of the resulting glucose, xylose, and other solubilized carbon components to hydrocarbon fuels in the gasoline, jet, and diesel fuel ranges (Biddy and Jones 2013).

Alcohol-to-Jet Fuel

Alcohol-to-jet fuel (ATJ) converts short carbon chain alcohols (such as methanol, ethanol, and butanol) to the longer C12/C16 alkanes of jet kerosene. The alcohol is produced conventionally (sugar/starch fermentation), thermochemically (e.g., gasification with upgrading), or through other pathways (industrial microbiology and algae). Several companies are exploring ATJ pathways, including Gevo and Cobalt. Gevo has developed a proprietary integrated fermentation technology (GIFT) consisting of a yeast biocatalyst that converts sugars into isobutanol. The alcohol is then converted into iso-paraffinic kerosene (IPK), a blendstock used in jet fuel, via additional reactions such as dehydration, oligomerization, hydrogenation, and distillation (Gevo 2011). Similarly, Cobalt has developed its own process for extracting sugars from biomass and converting them directly into bio n-butanol, a platform molecule for the production of a broad array of fuels and chemicals, including jet fuel (Cobalt 2013).

Capacity and Production

Current U.S. biodiesel production capacity is more than 1.8 billion gallons, with about 160 plants registered with the Environmental Protection Agency (EPA) under the RFS program (*Soybean Review* 2011). The industry struggled in 2010 due to increased feedstock prices and the expiration of a key

$1.00/gallon blender tax credit at the beginning of the year. According to the Energy Information Administration (EIA), biodiesel production decreased from 678 million gallons (Mgal) in 2008 to 343 Mgal in 2010 (EIA 2012a). The revised RFS (RFS2, initiated in July 2010) mandates the blending of 1 billion gallons per year of biomass-based diesel; biodiesel produced from soybean oil, animal fat, waste grease, and several other feedstocks qualifies for meeting this mandate. Driven by the RFS2 mandate and reinstatement of the $1/gallon blender tax credit, biodiesel production reached a record 1.1 billion gallons in 2011, kept that level in 2012, and it is expected to be higher in 2013 (EPA 2013a).

In comparison, the current U.S. renewable diesel and jet fuel production capacity is small, about 225 million gallons per year (Mgy) (*Biofuels Digest* 2012, KiOR 2013). The industry is just starting out and most of the facilities are at pilot/demonstration scale, under construction, or in planning phase. At the time of writing, there were three commercial facilities as described below.

Dynamic Fuels, a 50/50 joint venture between Syntroleum Corporation and Tyson Foods is located in Geismar, Louisiana. The facility has a capacity of 75 Mgy and became operational in November 2010. The plant uses animal fats, greases, and vegetable oils as feedstock. In the summer of 2011, Syntroleum announced that Dynamic Fuels achieved a record production of renewable fuels, about 87% of plant's capacity (Syntroleum 2011). High production level (71% of plant's capacity) was kept in 2012 as well (*The City Wire* 2013a). However, the plant was idled in November 2012 because of deteriorating market conditions and the partners opted for replacement of a catalyst in the facility that would increase production efficiency. That catalyst was installed in June 2013 but the plant still remains idle today, mainly because the partners can't reach amicable restart terms (*The City Wire* 2013b).

KiOR's production facility in Columbus, Mississippi became operational in early 2013. It uses woody biomass as feedstock and has 13 Mgy capacity (KiOR 2013). Given that the facility produces both renewable gasoline and diesel, it is unknown at this time the share of renewable diesel in the final output.

Diamond Green Diesel, a joint venture between Valero subsidiary Diamond Alternative Energy LLC and Darling International Inc. is located in Norco, Louisiana. The facility has a capacity of 137 Mgy and became operational in the summer of 2013. The plant uses recycled animal fat and used cooking oil as feedstock.

It is projected that other companies will come online within the next several years, although some may produce renewable gasoline instead of, or in addition to, renewable diesel and jet fuel (*Biofuels Digest* 2012).

PRODUCTION COST

This section presents production cost estimates for several technology pathways. It is important to note that the costs presented here are production costs rather than final selling prices at the pump.

Very little data are available publicly on the production costs for renewable diesel and jet fuel. Some techno-economic analyses (TEAs) have been conducted for certain technology pathways, and other data from specific companies can be extracted from S-1 filings when a company goes public.

In an effort to make the production costs reported here as comparable as possible, only recent sources (2008 to 2011) have been cited, and care was taken to exclude estimates with unusual assumptions such as optimistic feedstock price. However, the cost estimates are still not perfectly comparable. Variables such as plant size and co-product credits can have a significant impact on the overall production cost. Also, each pathway is at a different level of technology maturity, which contributes to the uncertainty of the cost estimates. None of the analyses cited address costs associated with carbon capture and sequestration (CCS), and none benefit from GHG emission credits.

Biodiesel

An analysis of soybean biodiesel (Tao and Aden 2009) reports a production cost of \$2.55 per gallon of biodiesel (\$2.48 gallon-of-gasoline-equivalent or gge) in 2007 dollars at a feedstock cost of \$0.30/lb of soybean oil. The feedstock cost is by far the most significant cost of biodiesel production, accounting for about 70% of the overall production cost, and 75%-95% of the overall operational cost. Other publications confirm the significance of feedstock cost, estimating it accounts for as much as 80% of the overall production cost (Yusuf et al. 2011). A literature review in Tao and Aden (2009) revealed overall production costs ranging from \$2.00-\$2.50 per gallon (\$1.94- \$2.43/gge), although the cost years of the data points in the literature review are not provided. Current biodiesel production facilities use a caustic soda-based catalyst, sodium methoxide, which complicates product clean up. Enzymatic transesterification using a lipase biocatalyst is an attractive alternative due to reduced wastewater treatments needs, easy glycerol recovery, and absence of side reactions—although the production costs are significantly higher (Jegannathan et al. 2011). An economic analysis employing an immobilized lipase catalyst resulted in a biodiesel production cost of $8.04 per gallon ($7.81/gge), although no cost year is provided.

Hydroprocessing of Lipid Feedstock

Pearlson et al. (2013) is the only publicly-available, peer-reviewed journal article that could be found containing production cost estimates for HEFA fuels. It models diesel and jet fuel production via hydroprocessing of soybean oil for three production capacity scenarios: 31, 61, and 100 Mgy. Maximizing the process for diesel production (jet fuel and LPG are co-products) results in cost per gallon ranging from $3.82 to $4.39/gal ($3.61 to $4.15/gge). Catalytically cracking a portion of the diesel range products was modeled to maximize jet fuel production (diesel and LPG are co-products), which resulted in a jet fuel cost per gallon of $4.09 to $4.69/gal ($3.81 to $4.37/gge). No cost year is provided, although input costs for the model, such as electric power, natural gas, and purchased hydrogen, are cited in 2010 dollars.

An estimate of renewable diesel production cost via the HEFA technology pathway can be gleaned from Syntroleum Corporation financial disclosure statements. Dynamic Fuels produces renewable diesel, along with naphtha and LPG co-products, from animal fats and yellow grease. According to Syntroleum Corporation financial statements (Syntroleum 2013), the feedstock was $3.62/gal of renewable diesel produced, and operating expenses were $1.03/gal of renewable diesel produced for the quarter ending September 30, 2012 (Syntroleum 2013). Cost per gallon due to capital costs is not provided, but the Dynamic Fuels website reports the capital cost of the 75 Mgy production facility at $150 million (Dynamic Fuels 2013). Assuming a 30-year project life and 10% discount rate, we can amortize a $150 million overnight capital cost over the total amount of fuel produced over the life of the project. The resulting capital cost is $0.22/gal. Adding the capital cost per gallon to the feedstock cost and operating cost results in a production cost of about $5/gge. It should be emphasized that this cost was not reported by Syntroleum Corporation, Tyson Foods, or Dynamic Fuels. It was estimated in this study based on feedstock and operating costs provided by Syntroleum, and total capital cost (provided by Dynamic Fuels) amortized using financing assumptions common in scoping studies such as the National Renewable Energy Laboratory (NREL) techno-economic reports cited elsewhere in this section.

Algae

Davis et. al. (2011) recently published modeled costs for the co-production of diesel and naphtha from photosynthetic algae. In this analysis, both open pond and photobioreactor technologies are examined. In general, the costs for finished (hydro-treated) blendstocks range from $9.84/gallon ($9.30/gge) for open-pond systems to $20.53/gallon ($19.39/gge) in 2007 dollars for photobioreactor systems.

Algal biofuel production is a nascent technology at commercial scale. Therefore, previous analyses apply inconsistent assumptions, making comparison of results difficult. Consequently, the Department of Energy's (DOE) Bioenergy Technology Office (BETO) hosted a workshop at the University of Arizona, Tucson, from November 30–December 1, 2011, for the purpose of harmonizing assumptions used in previous algal biofuel TEAs, life cycle assessments (LCAs), and resource assessments (RAs). Due to the favorable cost estimates of open-pond systems over photobioreactor systems, only open-pond systems were considered for the harmonization exercise. Applying the harmonized assumptions resulted in a cost of $19.60/gal ($18.52/gge) in 2007 dollars. The increase in cost above the Davis et al. (2011) estimate is due largely to the addition of pond liners, a reduction of the baseline algae production activity from 25 $g/m^2/day$ to 13.2 $g/m^2/day$, and accounting for location and seasonal variabilities (Davis et al. 2012).

Fischer Tropsch Diesel

Gasification and pyrolysis modeled costs have been most recently described in a series of papers from ConocoPhillips/Iowa State. Anex et al. (2010) estimate the production cost of diesel and gasoline produced via gasification of corn stover followed by FT synthesis at $4.50 to $5.00/gge in 2007 dollars, depending on the operating temperature of the gasifier. The National Energy Technology Laboratory (NETL 2009) estimates the production cost of diesel produced via gasification followed by FT synthesis as \$6.45/gallon (\$6.09/gge) in 2008 dollars. The NETL study assumes a 20% rate of return, whereas Anex et al. (2010) assume only a 10% rate of return.

Wright et al. (2008) explore distributed pyrolysis processing (on-farm and small co-op pyrolyzers) followed by centralized gasification and FT synthesis of the pyrolysis oil to FT liquids. The distributed processing scenario yields FT liquids for \$1.43 to \$1.56 per gge, although the analysis does not describe the composition of the FT liquid product. Additional separation and upgrading steps may be necessary. Additionally, transporting raw pyrolysis oil can be problematic due to its tendency toward phase separation. No cost year is provided for the estimate.

Non-Catalytic Fast Pyrolysis

Brown et al. (2011) perform a techno-economic analysis of a non-catalytic fast pyrolysis process that produces gasoline from corn stover for a selling price of \$2.96/gal in 2007 dollars. Biochar and pyrolysis gas are also produced, but are consumed in the overall process for heat generation. The final price is most sensitive to bio-oil yield, followed by feedstock price. The same research group analyzed two fast pyrolysis pathways, one with on-site hydrogen generation for fuel upgrading, the other relying on merchant hydrogen (Wright et al. 2010). Results show selling prices of \$3.09/gge for the on-site hydrogen production scenario and \$2.11/gge with purchased hydrogen. Selling prices in 2007 dollars for pioneer plants are \$6.55/gge for the on-site hydrogen production and \$3.41/gge with purchased hydrogen.

Catalytic Fast Pyrolysis

The National Academy of Sciences (NAS) reviewed cost estimates made by Wright et al. (2009) and catalytic fast pyrolysis company KiOR, based on KiOR's Form S-1 filing to the Securities and Exchange Commission (NAS 2011). Using its own feedstock cost and financing assumptions, the NAS updates the estimated selling prices of \$2.10/gge for Wright et al. (2009) and \$3.24/gge for the KiOR catalytic fast pyrolysis process, although most details of the analysis are not provided, and it is unclear what year the costs are indexed.

Hydropyrolysis

Marker et al. (2012) modeled an integrated hydropyrolysis and hydroconversion process for gasoline and diesel production, called IH^2, developed by the Gas Technology Institute (GTI). Biomass is pyrolyzed in the presence of hydrogen to gas and liquid products. The gas phase proceeds to a hydro-conversion stage, which removes oxygen, resulting in deoxygenated gasoline and diesel products. The production cost of the finished fuel is $1.60/gal in 2007 dollars. A gge production cost is not provided, although the product mix is reported as approximately 75% gasoline and 24% diesel, which results in a gge production cost of approximately $1.56 in 2007 dollars.

Biojet Fuel

A review of biojet fuel cost produced several estimates, but only one peer-reviewed journal article. Agusdinata et al. (2011) estimates the cost in 2007 dollars of producing biojet fuel via gasification and FT synthesis at $4.00/gge from corn stover, $5.50/gge for switchgrass, and about $5.80/gge for short rotation woody crops (SRWCs). Several estimates found in trade journals and industry blogs are consistent with this cost range of $4 - $6, although these sources are not reviewed and do not provide supporting material. Agusdinata et al. (2011) also estimates the cost of producing biojet fuel from algae in open ponds to be about $17/gge.

Renewable Identification Numbers and the Renewable Fuel Standard

The RFS program was created to ensure that transportation fuel sold in the United States contains a minimum volume of renewable fuel. The EPA is responsible for implementing the RFS regulations. The obligation to meet the minimum volume falls on fuel blenders.

Renewable Identification Numbers (RINs) are the mechanism used by the EPA to track volumes of renewable fuel and verify blenders are meeting minimum blending requirements. Each volume of renewable fuel produced has an RIN attached to it. Each fuel blender must acquire enough RINs to cover its share of the mandate. RINs are tradable and can be purchased and sold. Therefore, a blender can meet RFS requirements by actually blending the mandated amount of biofuels, or by purchasing RINs from other fuel blenders who blend more biofuels than required and have excess RINs (FAPRI-MU 2009).

The RFS and RIN trading system do not affect the production cost of renewable fuels, but are worth mentioning here because they can serve as economic drivers to make renewable fuels competitive with petroleum-derived fuels (NABC 2012a) and aid in initial market penetration.

DEMAND

Diesel

Distillate fuel oils, a category that includes diesel and heating oil, are a general classification for one of the fractions obtained from petroleum distillation. These oils are used in all sectors of the U.S. economy and rank second behind gasoline as the most consumed liquid fuels (EIA 2012c). Diesel fuel (often referred to as No. 2 diesel fuel) is defined by EIA as a fuel that has distillation temperatures of approximately 500°F at the 10% recovery point and no more than 640°F at the 90% recovery point; it meets the specifications defined in ASTM Specification D 975.[2] No. 1 diesel fuel is a light distillate fuel oil that has distillation temperatures of no more than 550°F at the 90% point and meets the specifications defined in ASTM Specification D 975. Both fuels are used in high-speed diesel engines found in trucks, buses, automobiles, and locomotives, as well as farm and construction equipment. No. 1 diesel exhibits a much lower cloud point than No. 2 diesel, so it is used neat, or it is blended with No. 2 in winter months. Heating oil (often referred to as No. 2 fuel oil) is used for domestic heating and for moderate capacity commercial/industrial buildings. EPA defines it as a distillate fuel oil that has a distillation temperature of up to 640°F at the 90% recovery point and it meets the specifications defined in ASTM Specification D 396.[2] No. 1 fuel oil is a light distillate fuel oil that has distillation temperatures of up to 400°F at the 10% recovery point and no more than 550°F at the 90% point; it meets the specifications defined in ASTM Specification D 396. It is used primarily as fuel for portable outdoor stoves and portable outdoor heaters and is commonly referred to as kerosene. More than three-quarters of distillate fuel sales are used primarily for transportation: on-highway (by trucks, buses, and automobiles); railroad; vessel bunkering; and construction, farm, and military equipment (Figure 1). About 6% of distillate sales are for residential heating purposes, concentrated during the winter months. Table 8 in the Appendix provides a detailed breakdown of the different types of fuels used in each sector, as well as their volumes during the period 2007-2012. On-highway motor vehicles consume about 64% of distillate fuel oils, namely diesel, with freight trucks using most of the fuel. The states of Texas and California are the largest consumers of diesel in the country, accounting for about 20% of all sales (Figure 2). Diesel consumption by on-highway motor vehicles follows population distribution: the top consuming states are also the most populated. Table 9 in the Appendix provides information on the historic use of No. 2 diesel fuel by state.

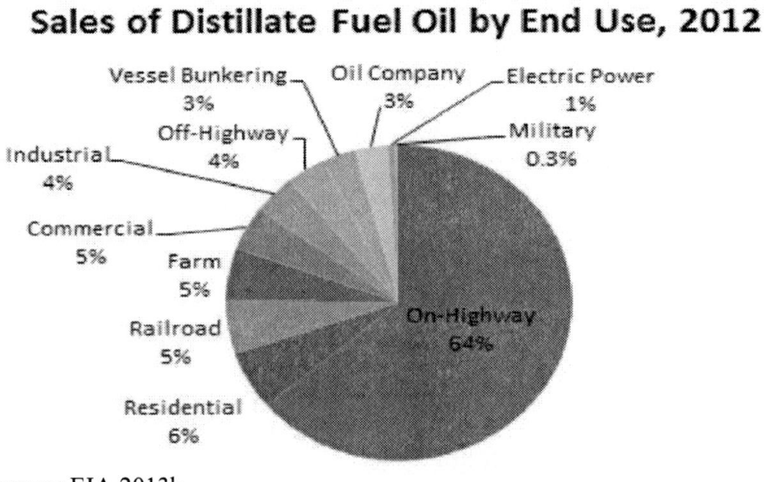

Data source: EIA 2013b.

Figure 1. Sales of distillate fuel oil by end use, 2012.

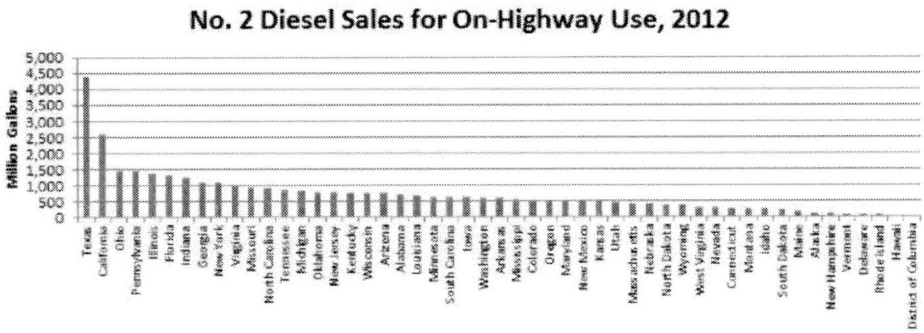

Data source: EIA 2013b.

Figure 2. No. 2 diesel sales for on-highway use by state, 2012.

Total distillate consumption was on the rise during the 1990s, and it reached a plateau during 2003-2007 (Figure 3). The growth was mainly due to increased use of diesel for transportation. From 2007-2009, low demand and a weak economy contributed to a downward trend. In 2010, this reversed to an upward trend, which experts believe is due to growth in manufacturing, usually associated with trucking demand.

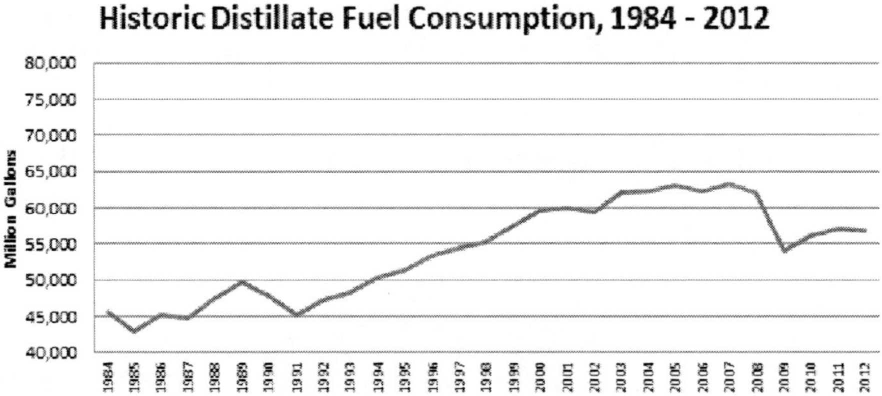

Data source: EIA 2013b.

Figure 3. Total U.S. distillate retail deliveries, 1984 – 2012.

Diesel consumption is projected to continue to grow, as illustrated in Figure 4. EIA notes that this growth results from both an increase in industrial output that leads to more fuel use by heavy trucks and an expansion of light-duty diesel vehicle sales to meet more stringent Corporate Average Fuel Economy (CAFE) standards.[3]

As shown in Figure 5, freight trucks are expected to continue to dominate diesel consumption. Freight rail fuel use is expected to increase slightly while the consumption of diesel by the remaining market segments (such as commercial light trucks, shipping, school buses, etc.) remains almost flat over the years (Figure 6).

An exception is the transit bus segment, which is projected to decrease its diesel use because of an increase in natural gas use. Historical diesel share of the transit bus segment declined relative to natural gas transit buses between 1995 and 2008, and this tradeoff is expected to continue.

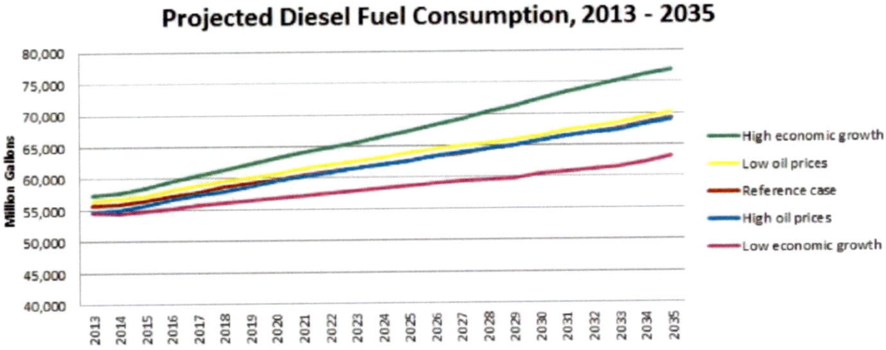

Data source: EIA 2011.

Figure 4. Projected diesel fuel consumption, 2013 – 2035.

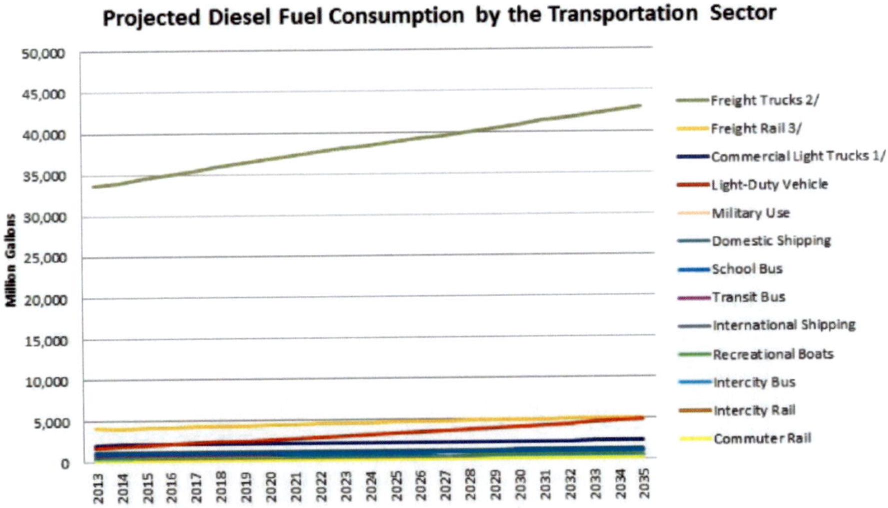

Data source: EIA 2011, Reference Case.

1/ Commercial trucks from 8,500 to 10,000 pounds.

2/ Does not include military distillate. Does not include commercial buses. 3/ Does not include passenger rail.

Figure 5. Projected diesel fuel consumption by the transportation sector.

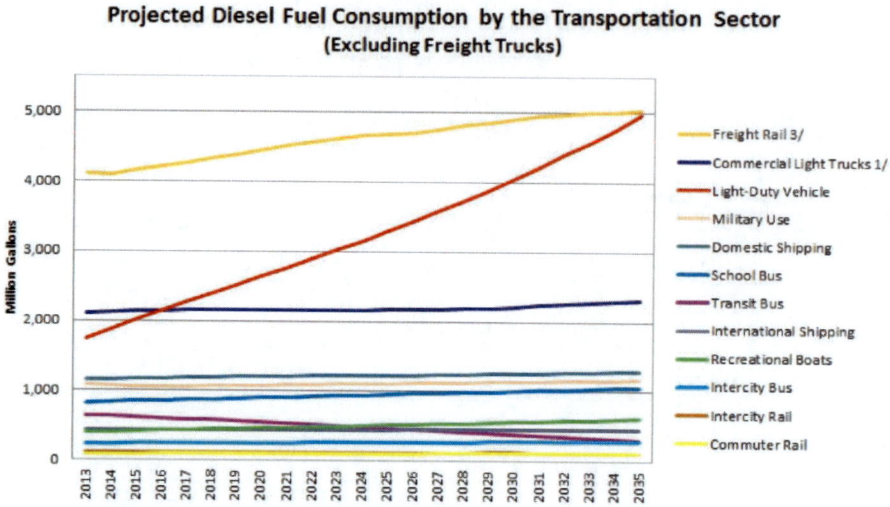

Data source: EIA 2011, Reference Case.
1/ Commercial trucks from 8,500 to 10,000 pounds. 3/ Does not include passenger rail.

Figure 6. Projected diesel fuel consumption by the transportation sector excluding freight trucks.

Biodiesel

Biodiesel consumption was on the rise from 2005-2009, plummeted during 2010 due to low demand resulting from the economic downturn and the expired $1.00 per gallon blender tax credit, and was back up in 2011 mainly in response to RFS2 and the reinstated tax credit (Figure 7). Biodiesel is distributed through fueling stations nationwide. Low-level biodiesel blends such as B2 and B5 are used safely in any compression-ignition engine designed to use diesel fuel. There are several hundred major fleets in the United States that use biodiesel. These include municipal bus fleets; national park trucks and buses; federal, state, and local government fleets; school buses; and many commercial businesses, such as public utilities and refuse haulers (Hart Energy Consulting 2010). Off-road applications (tractors, boats, and electrical generators) also use biodiesel. EIA projects an annual growth of 7% for biodiesel consumption by 2035 (EIA 2011).

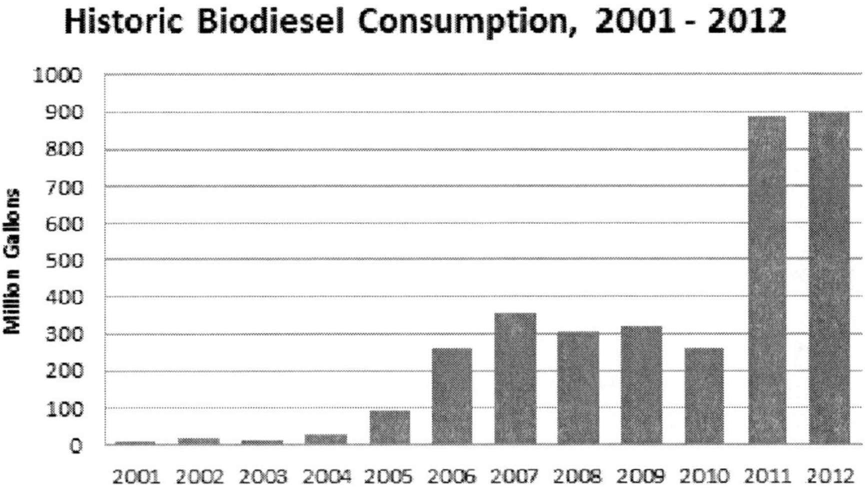

Data source: EIA 2013a.

Figure 7. Historic biodiesel consumption, 2001 – 2012.

Jet Fuel

Jet fuel is a type of aviation fuel designed for use in commercial and military aircrafts powered by gas-turbine engines. It is the third-most used fuel in the country after gasoline and diesel. EIA defines jet fuel as a kerosene-based product having a maximum distillation temperature of 400°F at the 10% recovery point and a final maximum boiling point of 572°F and meeting ASTM Specification D 1655 (JET A and JET A-1) and Military Specifications MIL-T-5624P and MILT-83133D (Grades JP-5 and JP-8).[4]

Jet fuel consumption was on the rise during the 1980s and 1990s and it reached a plateau during 2000 - 2007 (Figure 8). Since 2007, similar to diesel consumption, jet fuel consumption began a downward trend due to low demand and a weak economy. States consuming the most jet fuel include California, Texas, New Jersey, Florida, and Illinois (Figure 9); they are home to some of the busiest airports in the United States as well as large military bases.

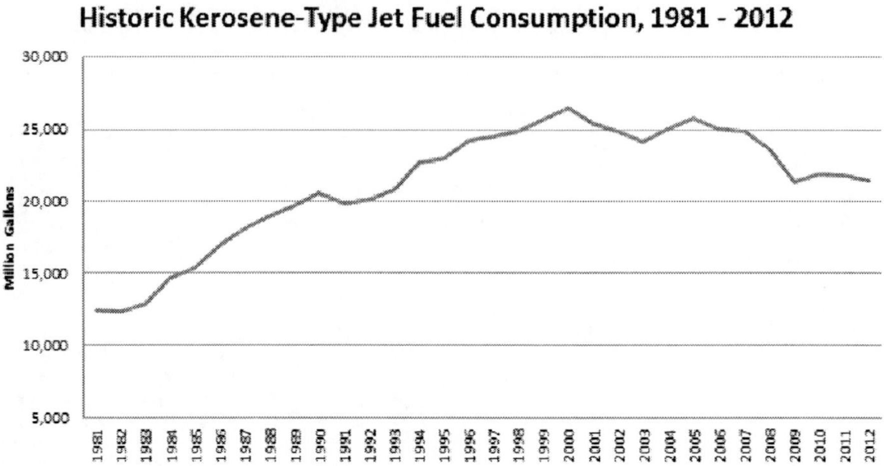

Data source: EIA 2013c.

Figure 8. Historic kerosene-type jet fuel consumption, 1981 – 2012.

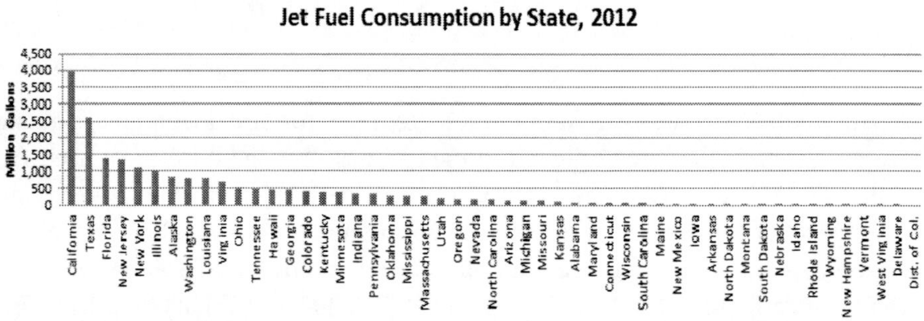

Data source: EIA 2013d.

Figure 9. Jet fuel consumption by state, 2012.

ETA projects that jet fuel consumption by commercial carriers will continue to grow over the next years, the rate of which will depend on economic growth and oil prices, whereas jet fuel consumption by the military will remain flat (Figure 10 and Figure 11).

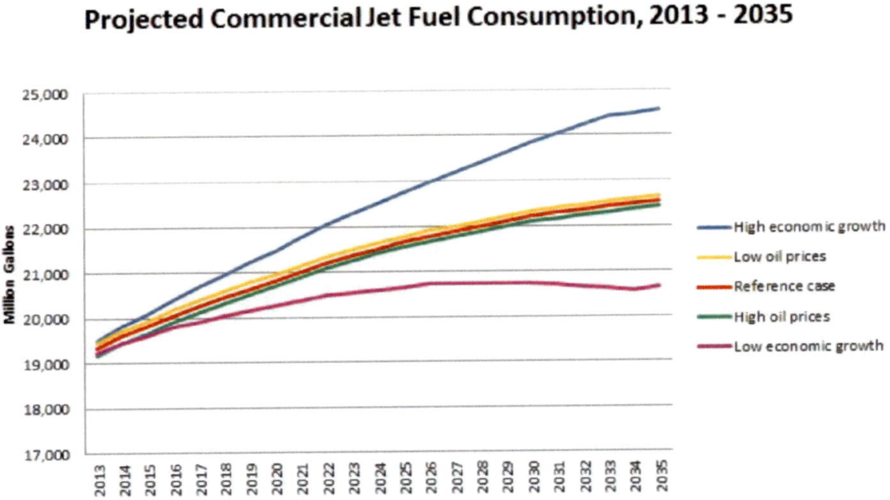

Data source: EIA 2011.

Figure 10. Projected commercial jet fuel consumption, 2013 – 2035.

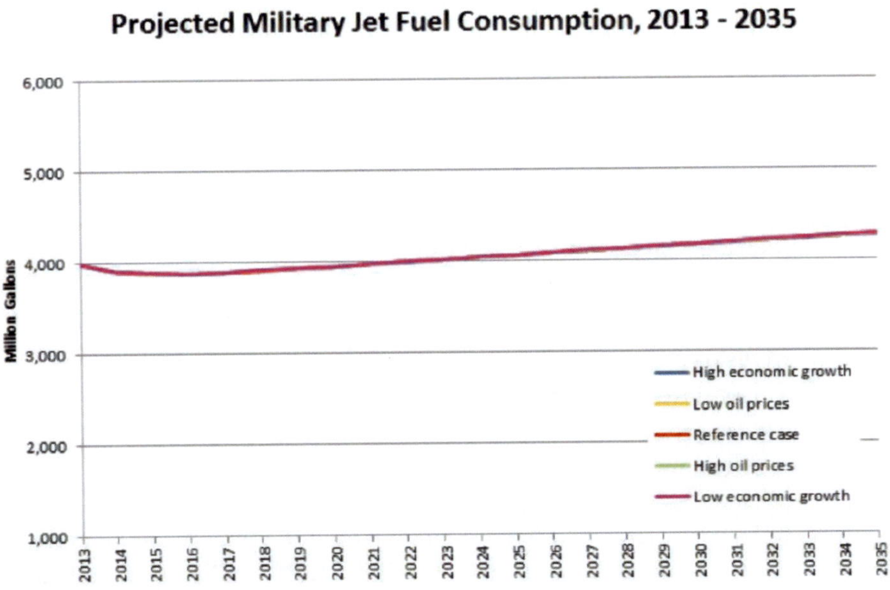

Data source: EIA 2011.

Figure 11. Projected military jet fuel consumption, 2013 – 2035.

FEEDSTOCK ASSESSMENT

Various biomass resources can be used for the production of diesel and jet fuel substitutes, as established earlier. These include lignocellulosic material such as wood waste, crop residues, and dedicated energy crops, as well as lipid feedstock such as vegetable and waste oils, animal fat, and algae. A recent study by the U.S. Department of Energy's Oak Ridge National Laboratory (U.S. DOE 2011), the Billion Ton Study (BTS), estimated the amount of currently available and potential biomass resources in the conterminous United States. The study included the following biomass categories:

1. Forest Biomass and Wood Waste

- Forest residues (logging residues and thinnings) from integrated forest operations from timberland
- Other removal residue[5]
- Thinnings from other forestland[6]
- Unused primary and secondary mill processing residues[7]
- Urban wood wastes (woody component of municipal solid waste [MSW] and construction and demolition [C&D] wood)
- Conventionally sourced wood[8]

2. Agricultural Biomass and Waste Resources

- Crop residues from the major grain-producing crops (corn, wheat, barley, oats, and sorghum)
- Secondary agricultural processing residues (sugarcane trash and bagasse, cotton gin trash and residues, soybean hulls, rice hulls and field residues, wheat dust and chaff, orchard and vineyard prunings)
- Waste or tertiary resources (e.g., manures, waste fats, and greases)

3. Energy Plantations (perennial grasses, trees, and annual crops).

The BTS makes use of POLYSYS, an agricultural policy model, and data from the United States Department of Agriculture (USDA) to estimate supply/cost curves for each feedstock by county for 2012-2030. Two scenarios are used: baseline and high yield. The baseline scenario assumes a continuation of USDA's 10-year forecast for the major food and forage crops, as well as a continuation in trends toward no-till and reduced cultivation.

Energy crop yields assume an annual increase of 1% due to experience in planting and additional R&D. Forest residues are estimated using resource cost analysis with data from the USDA's Forest Service. The high-yield scenario assumes a greater proportion of corn in reduced and no-till cultivation and increased corn yields to about double the current rate of annual increase. The energy crop productivity increases at 2%, 3%, and 4% annually, not only due to experience in planting, but also to more aggressive implementation of breeding and selection programs. No high yield scenario was evaluated for forest resources except for the woody crops.

The forest resources are available over a wider price range than the agricultural resources–$20 per dry ton or less to $100 per dry ton or less for forest resources vs. $40 per dry ton or less to $60 per dry ton or less for agricultural resources–with increasing quantities at higher prices. Over the estimated price range, forest resource quantities vary from about 33 to 142 million dry tons currently (Table 2). Due to data limitations, there is little estimated change over the next 20 years. Another study looked at the currently available forest resources in the country, regardless of price, and estimated that there are about 92 million dry tons of woody biomass available in the country today (Milbrandt 2005).[9] A study by NAS reports a similar amount of currently available woody biomass, about 110 million dry tons, and projects that about 124 million dry tons could be available in 2020 (NAS 2009).

Table 2. Summary of Potential Forest Biomass and Wood Wastes in the BTS, 2012

Feedstock ($ per dry ton)	<$20	<$30	<$40	<$60	<$80	<$100
	Million dry tons					
Other Removal Residue	4.4	12	12	12	12	12
Composite Operations	9.5	30	36	40	42	43
Without Federal Land	8.3	26	31	35	36	37
Treatment Thinnings, Other Forestland	0	0	0	3.2	6.4	6.4
Without Federal Land	0	0	0	1.8	3.6	3.6
Mill residue, unused primary	1.3	1.3	1.3	1.3	1.3	1.3
Mill residue, unused secondary	6.1	6.1	6.1	6.1	6.1	6.1
Urban Wood Waste – C & D	4.4	11	14	22	22	22
Urban Wood Waste – MSW	7.7	8.7	9.2	10	10	10
Conventional Pulpwood to Energy*	0	0	0	1.5	19	40
Total – All Land	**33**	**70**	**79**	**97**	**119**	**142**
Total – Without Federal Land	*32*	*66*	*75*	*90*	*111*	*133*

Source: U.S. DOE 2011.

Under the baseline scenario, agricultural resource quantities vary from about 59-162 million dry tons currently to about 126-265 million dry tons in 2030. About two-thirds of this quantity comes from crop residue and the rest is from various agricultural processing residues and wastes. Under the high yield scenario, the quantities vary from about 115-244 million dry tons currently to about 284-404 million dry tons in 2030 (Table 3). As a reference, Milbrandt (2005) reports about 157 million dry tons of crop residues available in 2002. This amount is one-third of the total crop residues, accounting for a portion that needs to be left on the field for soil protection, grazing, and other agricultural activities. NAS (2009) reports a lower amount of currently available agricultural resources[10]--about 106 million dry tons—and projects that about 148 million dry tons of agricultural resources could be available in 2020.

Table 3. Summary of Baseline and High Yield Scenarios in the BTS — Agricultural Residues and Waste Resources

	<$40 per dry ton				<$50 per dry ton				<$60 per dry ton			
Feedstock	2012	2017	2022	2030	2012	2017	2022	2030	2012	2017	2022	2030
	Million dry tons											
Baseline												
Corn	19	32	42	65	73	93	108	129	85	106	120	140
Wheat	6.7	7.8	9.1	12	18	22	26	31	23	26	31	36
Barley, Oats, Sorghum	1.0	1.3	1.6	2.9	2.4	2.5	2.4	3.6	2.8	2.7	2.6	3.7
Total primary residue	**27**	**41**	**52**	**80**	**94**	**117**	**136**	**164**	**111**	**135**	**154**	**180**
Secondary residues & wastes												
Rice field residue	6.5	6.9	7.4	8	6.5	6.9	7.4	8	6.5	6.9	7.4	8
Rice hulls	1.5	1.6	1.7	1.7	1.5	1.6	1.7	1.7	1.5	1.6	1.7	1.7
Cotton field residue	4.2	5.3	5.9	6.7	4.2	5.3	5.9	6.7	4.2	5.3	5.9	6.7
Cotton gin trash	1.4	1.6	1.7	1.8	1.4	1.6	1.7	1.8	1.4	1.6	1.7	1.8
Sugarcane residue	1.1	1.1	1.1	1.1	1.1	1.1	1.1	1.1	1.1	1.1	1.1	1.1
Orchard and vineyard prunings	5.7	5.6	5.5	5.5	5.7	5.6	5.5	5.5	5.7	5.6	5.5	5.5
Wheat dust	0.6	0.6	0.6	0.6	0.6	0.6	0.6	0.6	0.6	0.6	0.6	0.6
Animal manures	12	13	16	20	29	34	41	56	30	35	43	59
Animal fats	0	0	0	0	0	0	0	0	0	0	0	0
Total secondary residues & wastes	**33**	**36**	**40**	**46**	**50**	**56**	**65**	**82**	**51**	**58**	**67**	**84**
Total baseline	**59**	**77**	**92**	**126**	**143**	**174**	**201**	**245**	**162**	**192**	**221**	**265**
High-yield scenario												
Corn stover	71	132	157	221	143	200	228	264	153	209	234	271
Wheat Straw	9.8	12	13	16	60	35	38	42	35	39	42	46
Barley, Oats, Sorghum	1.5	1.5	1.4	1.7	3.6	3.4	2.8	3.1	4.0	3.6	2.9	3.0
Total primary residue	**83**	**146**	**171**	**238**	**176**	**239**	**269**	**309**	**193**	**252**	**279**	**320**
Total high-yield	**115**	**182**	**210**	**284**	**226**	**295**	**334**	**391**	**244**	**310**	**346**	**404**

Source: U.S. DOE 2011.

Energy crops considered in the BTS include switchgrass, miscanthus, sugarcane, sorghum, poplar, willow, eucalyptus, and southern pine. The study estimates that between 3.7 and 101 million dry tons of energy crops could be available under the baseline scenario in 2017 (Table 4). This wide range is due to uncertainty in crop yield and the rate of industry's development. The high yield scenario is simulated at three levels–2%, 3%, and 4% increase in annual crop productivity. Over the estimated price range, the quantity of energy crops varies between 13 and 180 million dry tons in 2017 to about 69 and 799 million dry tons in 2030. NAS (2009) reports that with current technologies and agricultural practices, about 104 million tons of energy crops could be produced today. Advanced technologies and practices could lead to increased production, potentially up to 164 million tons in 2020.

Table 4. Summary of Baseline and High Yield Scenario Availability of Energy Crops in the BTS

	<$40 per dry ton			<$50 per dry ton			<$60 per dry ton		
Feedstock	2017	2022	2030	2017	2022	2030	2017	2022	2030
Baseline scenario					*(Million dry tons)*				
Perennial grasses	3.0	12	30	41	77	129	90	188	255
Woody crops	0.0	0.0	0.1	0.9	40	67	5.7	84	126
Annual energy crops	0.7	1.8	4.2	3.8	7.3	14	5.0	10	19
Total	**3.7**	**14**	**34**	**46**	**124**	**210**	**101**	**282**	**400**
High-Yield (2% annual growth)									
Perennial grasses	11	43	57	67	152	239	122	253	319
Woody crops	0.0	0.1	4.2	1.9	78	127	10	145	207
Annual energy crops	1.6	4.1	7.4	5.5	8.7	12	6.9	11	15
Total	**13**	**47**	**69**	**75**	**239**	**378**	**139**	**409**	**540**
High-Yield (3% annual growth)									
Perennial grasses	24	71	107	85	213	329	138	296	390
Woody crops	0.0	1.5	43	9.3	101	186	14	168	251
Annual energy crops	2.4	6.6	11	6.2	10	14	8.0	12	18
Total	**26**	**79**	**162**	**101**	**324**	**520**	**160**	**476**	**658**
High-Yield (4% annual growth)									
Perennial grasses	35	100	202	106	270	406	154	338	462
Woody crops	0.1	5.3	45	12	118	199	16	212	315
Annual energy crops	3.4	9.0	14	6.8	11	18	9.4	14	22
Total	**39**	**114**	**261**	**124**	**399**	**622**	**180**	**564**	**799**

Source: U.S. DOE 2011.

FOG is another resource category for the production of diesel and jet fuel substitutes. The United States produced about 18 million tons of FOG in 2010 (Table 5), which could theoretically be converted to over 5 billion gallons of biodiesel or renewable diesel. Another 3.7 million tons of oils were imported (mainly edible oils such as canola, palm, and coconut). Edible oil imports have increased from 13% of domestic edible oil production in 1998 to 27% in 2010 as a result of the shift from soy to low polyunsaturated oils (because of negative health effects of trans-fatty acids and rising U.S. food demand). However, not all of these resources are available for energy use. Roughly 85% were used for edible (baking or frying fats, margarine, salad or cooking oil) and inedible products (fatty acids, animal feed, methyl esters, etc.). In addition, about 1.5 million tons of edible oils (mostly soy) and 900,000 tons of animal fats were exported in 2010 (USDA 2012; U.S. Census Bureau M311K). The 1.1 billion gallon biodiesel production in 2011 consumed over 4 million tons of FOG, which is a significant increase from the 1.1 million tons consumed in 2010.

Table 5. U.S. Production of Fats, Oils, and Greases in 2010

Sources	Production ('000 tons)
*Oils**	
Corn	1,258
Cottonseed	408
Peanut	82
Canola	576
Safflower	32
Soybean	9,518
Sunflower	314
Tall oil, crude**	672
Vegetable foot**	178
Fats	
Lard*	425
Edible tallow*	913
Inedible tallow**	1,650
Poultry fat**	709
Yellow grease**	702
Other grease**	660
Total Current Supply	18,092
*Oils and Fats Import**	3,687

* Data Source: USDA, Oil Crops Yearbook 2011.

** Data Source: U.S. Census Bureau, M311K.

Algae are a potential aquatic oil crop, but may also yield carbohydrates that can be converted to sugar. This feedstock has received increased attention in recent years. Given the right resources– suitable climate, availability of water, CO2 and other nutrients – algal oil productivity can be quite high. A recent study suggests that under current technology, microalgae have the potential to generate 58 billion gallons of oil per year, equivalent to 48% of current U.S. petroleum imports for transportation (Wigmosta et al. 2011). The authors emphasize that this level of production requires 5.5% of the land area in the conterminous U.S. and nearly three times the water currently used for irrigated agriculture. The intensive water use for growing algae can be addressed in a sustainable manner by using low-quality water with few competing uses, such as brackish/saline groundwater, "co-produced water" from oil and natural gas wells, and wastewater discharged from domestic, industrial, and agricultural activities. Therefore, algal technology need not put additional demand on freshwater supplies.

Figure 12 illustrates the algal oil productivity at different locations in the United States. The pattern shows a strong linkage to climate and topography– locations with warm temperatures and flat terrain are most productive. The southern portions of the country exhibit the highest production rates ranging from 6,000 to 8,221 liters/ha/year (between 540-740 gallons/acre/year) of potential biofuel production. These areas are characterized with relatively warm temperatures year-round and longer hours of solar insolation in comparison to northern locations. Locations in the north and at higher elevations exhibit the lowest production rates, ranging from 2,291 to 4,000 liters/ha/year (206-360 gallons/acre/year). A long winter season in these areas contributes to a shorter growing season.

Table 6 summarizes the biomass resource potential outlined above. Depending on the conversion pathway, different yields of biofuels can be achieved. Using current conversion technologies for lignocellulosic biomass, about 43.3 gallons of renewable diesel/jet fuel could be produced per dry ton via gasification/FT technology (Davis 2009) and about 65 gallons per dry ton via fast pyrolysis (PNNL 2009). For example, forest resources alone could produce about 1.4 - 6 billion gallons of renewable diesel via gasification/FT technology which could displace between 4% and 17% of current diesel consumption on highways (about 36 billion gallons of diesel fuel was used in 2011; see Figure 1 and Figure 3). Using fast pyrolysis, forest resources could yield even more "drop-in" fuels and displace between 6% and 26% of current diesel consumption. Renewable diesel and jet fuel yields from lignocellulosic biomass via biochemical conversion pathways are not readily available at this time.

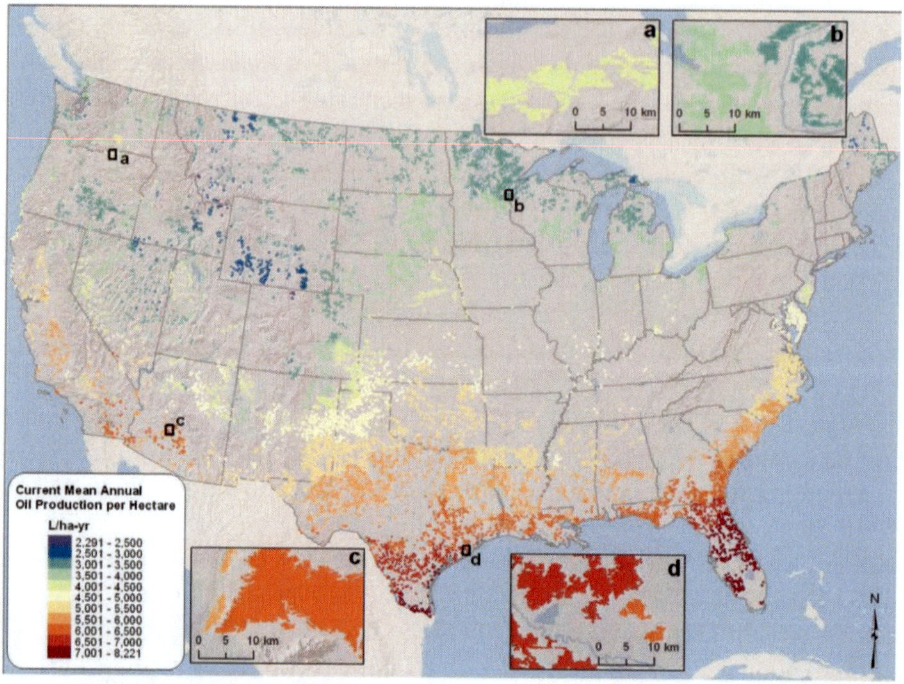

Source: Wigmosta et al. 2011.

Figure 12. Mean annual algal oil production using current technology.

On average, about 300 gallons of biodiesel, using a transesterification process, could be produced per ton of FOG. Thus, the 3.8 million tons of currently available FOG could yield about 1.1 billion gallons of biodiesel, which could displace about 3% of current diesel consumption on highways in the United States. About 245 gallons of "drop-in" fuel could be produced per ton of triacylglycerol (TAG) oil via hydroprocessing (Davis et al. 2012). Thus, about 931 million gallons of renewable diesel could be produced from the existing FOG in the United States, which could displace about 2.5% of current diesel consumption on highways in the country.

It is unrealistic to assume that all of the biomass resources could be used for the production of diesel and jet fuel substitutes. There are other competing uses for these resources. For lignocellulosic biomass, these include the production of ethanol, renewable gasoline, chemicals and allied products, as well as power generation. For FOG, competing industries include the production of edible (baking/frying fats, cooking oil, etc.) and inedible (lubricants, paints, plastics, etc.) products. However, even if a small portion of the current biomass potential is dedicated to diesel and jet fuel substitutes, they can still play a role in diversifying the country's energy portfolio. This role would be more significant if the projected future potential is realized and feedstock, such as dedicated energy crops and algae, become commercially available.

Table 6. Total Biomass Resource Potential

Feedstock	Current Potential (million tons)	Future Potential in 2022 (million tons)
Forest resources	33-142	34-150
Crop residues	59-162	92-221
Energy crops	n/a	14-282
Fats, Oils, and Greases*	3.8	4.1
Algae**	n/a	400

* Assumes 21% of FOG [about 85% of total FOG (18 million tons) is currently used for edible and inedible products, of which 6% is used in methyl esters (biodiesel)]. Future potential is estimated using population projections from the U.S. Census Bureau in 2022 (0.93 percent change) and applying the same ratio (21% of total).

** Assumes 1/5 of projected future potential.

DISCUSSION

The Congressionally mandated RFS2 goal is to use at least 36 billion gallons of biomass-based transportation fuels by 2022 (Figure 13). Of the total 36 billion gallons, 15 billion gallons are projected to come from conventional biofuel sources, such as corn ethanol, and the remaining 21 billion gallons from advanced biofuels divided into three categories: cellulosic biofuels, biomass-based diesel, and other advanced biofuels (EPA 2009, Table V.A. 2-1). Sixteen billion gallons are required to come from cellulosic biofuels (fuels, not necessarily ethanol, made from lignocellulosic biomass that also reduce GHG emissions by at least 60% relative to the gasoline or diesel fuel they displace). The contribution of biomass-based diesel to this goal can be no less than 1 billion gallons[11]: 0.81 billion gallons are projected to come from biodiesel and 0.19 billion gallons from renewable diesel. An additional 4 billion gallons of other advanced biofuels (any renewable fuel other than ethanol derived from corn that achieves 50% GHG emissions) is also mandated by RFS2.

Biomass-based diesel and jet fuel are considered to be advanced biofuels and can meet several RFS criteria. For example, renewable diesel derived from lignocellulosic biomass falls within the biomass-based diesel and cellulosic biofuels categories while jet fuel substitutes from biomass can meet the cellulosic biofuels and other advanced biofuels standards.

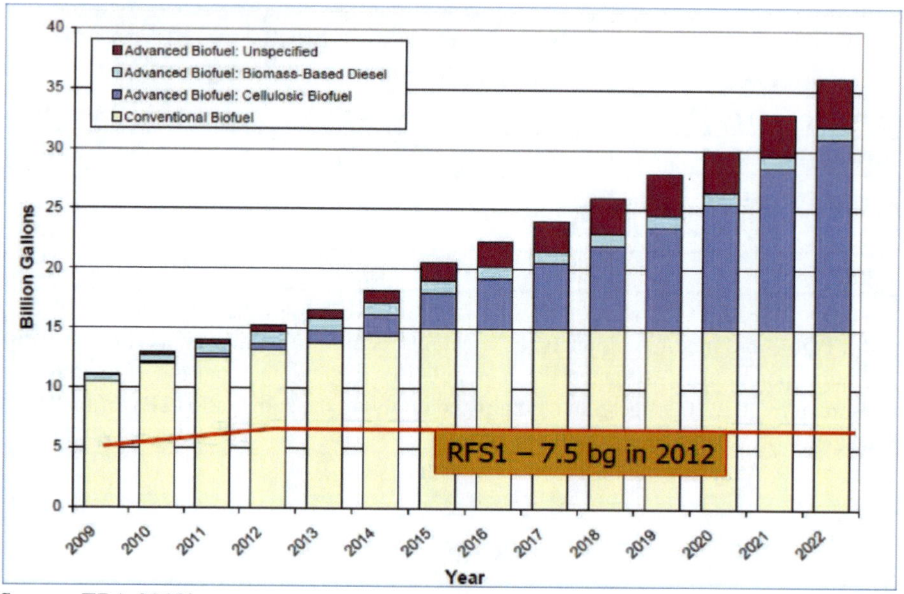

Source: EPA 2010b.

Figure 13. Renewable fuel volume requirements for RFS2 under EISA.

During 2011 and 2012, the biodiesel industry proved that it has the ability to fulfill the entire biomass-based diesel volume requirements under RFS2. The mandate calls for at least 1 billion gallons of biomass-based diesel annually and biodiesel production was reported at 1.1 billion gallons during these two years. Production in 2013 is forecasted to be even higher. In addition to the RFS mandate, another factor for the industry's growth during the past three years was the extended tax credit. This key $1.00-per-gallon blender tax credit is scheduled to expire on December 31, 2013 and it is unknown at this point whether or not it will be extended. The industry could face significant setbacks if extension does not occur. In addition, EPA proposes to keep the biomass-based diesel standard at 1.28 billion gallons next year, the same volume set for 2013, which could further jeopardize the industry's further development (EPA 2013c). The next several months will reveal the biodiesel industry's future directions as these two legislative pieces get resolved.

Biofuels Digest (2012) projects that "drop-in" fuels capacity (inducing renewable diesel and jet fuel) will reach about 1 billion gallons in 2017. This projection is based primarily on companies' statements and media releases on planned facilities. Whether or not this capacity level is reached will depend on many factors, including commercialization of proven technologies (namely gasification, FT, and pyrolysis), success of trial technologies, feedstock availability, production cost, and diesel/jet fuel demand. As mentioned earlier, currently there are three commercial plants in the United States producing renewable diesel, although one of them (KiOR) produces other renewable fuels as well such as renewable gasoline. At present, Dynamic Fuels is idle and it is unknown when the facility will resume operation again. The other two plants, KiOR and Diamond Green Diesel are in an initial phase and thus it is difficult to evaluate the performance and economics of these plants. Neste Oil is the world's largest renewable diesel producer with about 572 MGY of installed capacity worldwide (*Biofuels Digest* 2012). It operates abroad in Finland, Singapore, and the Netherlands. While Neste Oil is using hydrotreating, a common practice in petroleum refineries, other technologies such as pyrolysis (employed by KiOR) and FT (utilized by Syntroleum) have yet to demonstrate their economic viability. The next few years will likely answer many questions about the feasibility and profitability of producing renewable diesel and jet fuel through the various conversion pathways.

Based on current statistics, and as proven by the biodiesel industry, there is enough lipid feedstock to support the RFS2 mandate. Today, roughly half of the biodiesel in the United States is produced from soybean oil. The remaining portion consists of animal fat, used cooking oil, canola, and some other minor feedstocks such as cottonseed oil, tall oil fatty acids, and corn oil recovered from corn ethanol dry mills. The Food and Agricultural Policy Research Institute (FAPRI) projects that soybean oil production in the United States will continue to rise over the next decade, from about 9.5 million tons today to about 11 million tons in 2022 (FAPRI-ISU 2012). One of the drivers for this increase is the anticipated volume required to meet the biofuels quota under the RFS2. Even if half of the projected soybean potential is used for biodiesel/renewable diesel production it would be sufficient to meet the mandate. Higher output could be achieved through advanced technologies for increasing oil supply (increasing crop yield and seeds' oil content, as well as recovering sewage and trap greases) and production of new feedstock such as oil from camelina (*Camelina sativa* L.) and field pennycress (*Thlaspi arvense*). Moreover, if algal oil becomes commercially available in the next 5-10 years, the biodiesel and renewable diesel production potential would increase substantially. With regard to the lignocellulosic feedstock, enough material is projected to be available to support the RFS mandate of 21 billion gallons from advanced biofuels. Crop and forest residues in 2022 are projected to be between 126 - 371 million tons. Assuming a conversion via fast pyrolysis (with an average yield of 65 gallons/dry ton), this amount could yield about 8-24 billion gallons of biomass-based diesel/jet fuel. This potential could be larger if the conversion technologies achieve higher yield and if additional feedstock such as dedicated energy crops become available. The fuel produced from these resources would qualify under any of the advanced biofuels categories: cellulosic biofuels, biomass-based diesel, or other advanced biofuels. However, there will be competition for lignocellulosic feedstock with the ethanol industry and renewable gasoline producers to meet the RFS mandate, so it is unclear what share the renewable diesel/jet fuel would have in the total biofuels contribution. Ultimately, it will depend on the rate of commercialization of these technologies as well as the transportation market demands. The costs for producing renewable diesel and jet fuel are not well known and include a high degree of uncertainty. Hydroprocessing of FOG is being done commercially by Dynamic Fuels, Diamond Green Diesel, and internationally by Neste Oil. The KiOR technology is at an initial commercial scale. Other technology routes are not yet commercial and publicized cost estimates vary widely. As these technology pathways mature and become more widespread, more specific information regarding their economics will be available, which would enable a more detailed analysis and performance comparison. Based on Syntroleum Corporation financial disclosure statements, we estimate the production cost for renewable diesel by Dynamic Fuels to be about \$5/gge as of September 2012. As a reference, the average production cost of petroleum-derived diesel in the United States is about \$2.87 (\$2.59/gge) as of October 2013 (EIA 2013e).[12] As another reference, production cost for biodiesel is in the range of \$2.00-\$2.50 per gallon (\$1.94-\$2.43/gge). However, it should be noted that this range was documented in 2009; thus it is likely that biodiesel production cost is higher today given the increase in soybean and other relevant feedstock prices during the past several years (USDA 2013). As a reminder, this is the cost of production (feedstock and processing), not the price paid at the pump by the consumer. The price at the pump also includes distribution expenses, taxes, etc. Biodiesel faces some technical limitations which may direct future industry decisions toward renewable diesel. While the lower-level biodiesel blends (B20 and below) can be used in traditional diesel vehicles without engine modifications, higher-level blends may require engine modifications and other usage considerations. More importantly, biodiesel is currently transported from the production sites to petroleum terminals where it is blended with petroleum fuels, via truck or rail and occasionally, by barge. These modes are much more expensive than transportation by pipeline, which is used for most petroleum fuels. The potential for biodiesel to contaminate jet fuel is preventing widespread pipeline transport. Hydrocarbon renewable diesel is likely to be easily transported in existing pipelines already utilized for petroleum-based fuels.

Renewable diesel and jet fuels produced by hydroisomerization of lipid feedstocks, gasification followed by FT synthesis, or biochemically using the approaches by Amyris, LS9, and other companies will consist entirely of paraffinic hydrocarbons (also known as alkanes). These are completely miscible with conventional fuels and, from a purely combustion standpoint, they could be blended at any level or even used in neat form as long as they meet the requirements of the respective fuel standard specifications (ASTM D975 for diesel fuel and ASTM D1655 for jet fuel). For both diesel and jet applications, it is likely that fuel additives will be required to meet specifications for lubricity, conductivity, stability, and other properties. All of these additives are also commonly used in petroleum-derived fuels. However, fuel system elastomeric components for both ground transport and aviation have been conditioned over time in a significant level of aromatic compounds. When exposed to a zero aromatic fuel such as the renewable diesel and jet fuels described above, aromatics can be extracted from the elastomers, causing them to shrink and leading to fuel system leaks. To avoid this scenario in aviation applications, ASTM International has developed standard D7566 "Standard Specification for Aviation Turbine Fuel Containing Synthesized Hydrocarbons," which allows no more than 50% volume paraffinic kerosene in jet fuel. The blended fuel is also required to contain a minimum of 8% volume aromatics. A similar situation could be envisioned for elastomeric components in diesel engine fuel systems, but to date, ASTM has not specified a minimum required aromatic level or a maximum allowable blend level of paraffinic components.

Other renewable diesel fuels, such as those produced by fast pyrolysis and related thermochemical processes, will likely contain significant levels of aromatics, and perhaps low levels of oxygenates. More data on the composition of these fuels will be required in order to assess any possible limit on blend level. For widespread application, biodiesel blending is limited to

20% volume due to lack of engine manufacturer approval and limited data on compatibility with infrastructure for higher blends.

As noted earlier, on-highway motor vehicles consume about 64% of the diesel fuel in the country, with freight trucks being the largest consumer in this category. Therefore, the trucking industry could provide a strong business opportunity for biomass-based diesel producers. The American Trucking Association (ATA) is supportive of alternative fuels, evident from the fact that it currently supports biodiesel use in blends up to 5%. The organization highlights that renewable diesel has not been subjected to rigorous on-road fleet testing; however, preliminary information indicates that renewable diesel may have advantages over biodiesel for the end-user (ATA 2012). These advantages may include a higher energy content and better cold weather performance compared to biodiesel. Regardless, for any biomass-derived diesel to be a success among the trucking companies, the fuel price has to be equal to or less than petroleum-based diesel. Table 7 illustrates the top 25 U.S. trucking companies in 2012. By far, the two major package delivery companies UPS and FedEx lead the ranking in revenue. The e-commerce boom in recent years is the main contributor to the success of these companies.

As illustrated earlier, diesel consumption by light-duty vehicles in the United States is relatively low, but it is projected to double over the next decade. The fuel is not as popular among U.S. passenger-car drivers as it is in Europe, but diesel vehicle sales show consistent double-digit increases over the last two years (*PR Newswire* 2012). Whether diesel cars remain niche vehicles in the U.S. market or enter the mainstream will depend on government policies, consumer demand, and fuel prices. The recently issued new CAFE standard of 54.5 mpg by 2025 is expected to have a positive effect on future clean diesel car sales. The Diesel Technology Forum (DTF), a non-profit educational organization dedicated to raising awareness about the economic importance and essential uses of diesel engines, issued a statement welcoming the new standards: “Because clean diesel autos are 20 to 40 percent more efficient than gasoline vehicles, diesel will be a major player in the nation’s effort to achieve the new mileage standards” (*Time* 2012, 8). Demand for gasoline is higher in the United States and fuel taxes favor gasoline, which makes gas less expensive. It is exactly the opposite in Europe where the fuel tax structure favors diesel (Figure 14). The U.S. diesel passenger vehicle market may expand as consumers are drawn more to diesels because they’re economical, offering greater fuel efficiency than comparable gas-powered cars. Moreover, if the U.S. fuel prices take off more abruptly than analysts predict, we could see a deeper penetration of diesel vehicles.

Table 7. Top 25 U.S. Trucking Companies in 2012

2012 Rank	Parent Company	Freight Service	Subsidiary Portfolio/Services & Comments	Public Private	2011 Annual Revenue ($million)	2012 Annual Revenue ($million)	2011-2012 Percent Change
1	UPS	Parcel	*UPS Ground, UPS Freight*	Public	$24,752	$25,692	3.80%
2	FedEx	Parcel	*FedEx Ground, FedEx Freight, FedEx Custom Critical*	Public	$14,149	$15,416	9.00%
3	J.B. Hunt Transport Services	TL	*Truckload, Dedicated Contract Services, Integrated Capacity Solutions, Intermodal*	Public	$4,527	$5,055	11.70%
4	YRC Worldwide	LTL	*YRC Freight, YRC Regional*	Public	$4,839	$4,851	-0.40%
5	Con-way	LTL	*Con-way Freight, Con-way Truckload*	Public	$3,729	$3,898	4.50%
6	Swift Transportation	TL	*Truckload, Dedicated, Intermodal*	Public	$3,334	$3,493	4.80%
7	Schneider National	TL	*Schneider National, Schneider National Bulk Carriers, Schneider Intermodal*	Private	$3,170	$3,262	2.90%
8	Landstar System	TL	*BCO, TBC, Intermodal*	Public	$2,739	$2,952	5.90%
9	Old Dominion Freight Line	LTL	*Fastest-growing LTL carrier in Top 50 List*	Public	$1,883	$2,110	12.10%
10	Werner Enterprises	TL	*One-way Truckload, Dedicated, Cross-Border*	Public	$2,003	$2,036	1.70%
11	Arkansas Best	LTL	*ABF Freight System, Truck Brokerage, Panther Expedited*	Public	$1,756	$1,900	8.20%
12	U.S. Xpress Enterprises	TL	*Arnold Transportation subsidiary merged with LinkAmerica in December 2012*	Private	$1,670	$1,730	5.40%
13	Estes Express Lines	LTL	*Largest privately held LTL carrier*	Private	$1,640	$1,754	6.90%
14	Prime	TL	*Largest temperature-controlled carrier*	Private	$1,236	$1,372	13.00%
15	R & L Carriers	LTL	*R+L Carriers, R+L Truckload Services*	Private	$1,207	$1,250	3.60%
16	C.R. England	TL	*National, Regional, Dedicated, Intermodal, Mexico*	Private	$1,005	$1,146	14.10%
17	Greatwide Logistics	TL	*Merged with Cardinal Logistics in February 2013*	Private	$1,046	$1,105	5.60%
18	Saia	LTL	*Saia Motor Freight Line*	Public	$1,030	$1,099	6.60%
19	Kenan Advantage Group	TL	*Acquired Jack B. Kelley in September 2011*	Private	$988	$1,091	10.40%
20	CRST International	TL	*Acquired Specialized Transportation (STI) in July 2011*	Private	$846	$1,051	25.40%
21	Crete Carrier	TL	*Crete Carriers, Shaffer Trucking, Hunt Transportation*	Private	$942	$999	6.10%
22	Roadrunner Transportation	LTL	*Completed 6 acquisitions in 2012. Largest year-over-year growth in the Top 50 Trucking list*	Public	$758	$958	28.60%
23	Knight Transportation	TL	*Dry Van, Refrigerated, Brokerage, Port & Rail Services, Intermodal*	Public	$856	$936	8.10%
24	Southeastern Freight Lines	LTL	*Launched brokerage service in January 2011*	Private	$820	$875	6.60%
25	Averitt Express	LTL	*Revenue growth driven by LTL division*	Private	$759	$739	2.60%

Data source: JOC 2013.

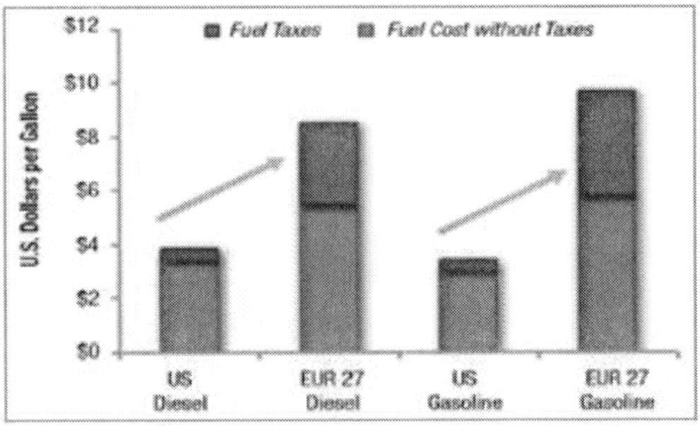

Fuel taxes for the United States include $0.184/gal for gasoline and $0.244/gal for diesel.

Source: U.S. DOE 2010.

Figure 14. Comparison of fuel prices and taxes in the United States and Europe, 2007.

Other transportation segments that could provide business opportunities for the biomass-based diesel industry are railroad; vessel bunkering; and construction, farm, and military equipment. While overall these segments do not consume as much diesel as freight trucks, they could still play a role at local levels. Tables 10-14 in the Appendix illustrate these sectors' diesel consumption by state. The rail network spreads throughout all states but certain areas such as Texas, California, Oklahoma, and Ohio report significantly more diesel consumption by the rail industry. The sector is strong in Texas and California due to a concentration of industrial and agricultural activities, in Oklahoma due to the state's somewhat central location, and in Ohio due to movements between eastern locations and markets in the Midwest. Diesel use by vessel bunkers is naturally high in coastal areas, while highly populated states (namely California and Texas) stand out as the largest consumers of diesel by the construction industry. About half of the country's diesel consumption in farm machinery is concentrated in the Midwest; large states with strong agricultural activities such as California and Texas also report high diesel use by this sector. As noted earlier, the military is not a heavy consumer of diesel; most activities are concentrated in states with large military bases such as Washington, Texas, North Carolina, and California.

Jet fuel is forming as a large and profitable market for the renewable fuels industry. Success in this area could stimulate a significant increase in the production of biofuels and associated feedstock. Successful test flights by commercial and military aircraft over the past few years led to the adoption of specification D7566 by the ASTM International in July 2011, which enables the use of 50% by volume paraffinic renewable fuels with conventional jet fuel. The blending level is restricted to 50% to ensure that the final blend contains a minimum level of aromatic compounds to prevent shrinkage of aged elastomer seals and subsequent fuel leakage. Since then, renewable jet fuel blends have been used by a number of airlines. The German airline Lufthansa began regularly scheduled commercial flights using Neste Oil's NExBTL renewable aviation fuel derived via hydrotreating vegetable oils and animal fats. This made Lufthansa the world's first airline to utilize biofuel in daily flight operations and Neste Oil the first company to provide the biofuel to be used on regularly scheduled commercial flights (Neste Oil 2012). In November 2011, Continental Airlines Flight 1403 from Houston to Chicago was the first U.S. commercial flight to run on a biofuel blend derived partially from genetically modified algae, which was provided by Solazyme. The parent company, United Continental Holdings Inc., plans to buy 20 million gallons per year of algae-derived biofuel made by Solazyme (*Seeking Alpha* 2012). Shortly after, Alaska Airlines and its sister airline Horizon Air started flights using a biofuel blend made from recycled cooking oil. More U.S. airlines are expected to join the effort to fly with renewable jet fuel in the following years.

The U.S. military is a promising client for renewable fuels given the fact that the Department of Defense (DoD) is the single largest consumer of petroleum in the country. The Defense Logistics Agency reports that it purchased about 4.7 billion gallons of fuel in FY 2012, of which 75% represented jet fuel (DLA 2013). Considering the approval of a 50% biofuels blend, this consumption represents a substantial potential demand for biofuels from the U.S. military in the near term. Among all services, the Air Force consumes the most energy and uses more than 2 billion gallons of aviation fuel annually. By 2016, it plans for half of its domestically purchased aviation fuel to be derived from renewable resources. Similarly, the Navy's goal is to use 50% of its energy consumption from alternative sources by 2020. This goal came as a result from successfully testing an F/A-18 fighter jet and helicopters on biofuel blends. The military plays a very strategic role in the development and commercialization of advanced biofuels. It can afford research and testing of alternative fuels to determine their viability more easily than the private sector, which is somewhat limited by concerns about returns on investment. Also, it can purchase large volumes of fuel to create demand, which could lead to increased production and lower prices. Moreover, early adoption of advanced fuel technologies by the DoD provides certainty to investors that there will be a market for these products.

CONCLUSION

Our examination of the status, opportunities, and feasibility for biomass-based diesel and jet fuel in the United States leads to the following findings:

1) It is technically feasible to produce biomass-derived diesel and jet fuel substitutes in the United States. Many conversion technology options exist. Some are commercially available or in demonstrational stage; others are still in the research and development phase.
2) Biodiesel, consisting of FAME produced from lipids, is currently the predominant form of biomass-based diesel. Production reached a record 1.1 billion gallons in 2011, kept that level in 2012, and is expected to be higher in 2013. Biodiesel blends cannot yet be considered fully "drop-in" fuels because they cannot be transported in all petroleum product pipelines. For pipelines that transport jet fuel, there is a concern that the jet fuel will be contaminated with biodiesel,

making it unsuitable for use. Ongoing research aims to determine what, if any, level of FAME can be tolerated in jet fuel.

3) In comparison, the current U.S. renewable diesel and jet fuel production capacity is small, about 225 million gallons per year. These fuels can be produced from various biomass resources and through several different approaches which all target hydrocarbon products that are similar to petroleum fuels in chemical makeup, and therefore may be considered "drop-in" fuels. It is anticipated that, as "drop-in" fuels, they can be blended with petroleum diesel/jet fuel at high levels, or possibly used in neat form.
   - Renewable diesel is produced at commercial scale primarily by hydroisomerization of lipid feedstock. Currently, there are two commercial facilities utilizing this process: Dynamic Fuels (Geismar, Louisiana) and Diamond Green Diesel (Norco, Louisiana).
   - A number of processes are under development for production of renewable diesel from biomass-derived sugars (corn, sugarcane, and sorghum, as well as sugars from thermochemical or biochemical depolymerization of cellulose and hemicellulose).
   - Processes are also being developed for direct conversion of lignocellulosic biomass by fast pyrolysis, gasification, and other thermochemical means. The first commercial plant is operated by KiOR in Columbus, Mississippi and became operational in early 2013.
4) The costs for producing renewable diesel and jet fuel are not well known and involve a high degree of uncertainty. The process economics for these fuels is highly dependent upon the cost of the feedstock, similar to biodiesel. Additionally, variables such as plant size and co-product credits can have a significant impact on the overall production cost. Hydroisomerization of lipids is performed commercially by Dynamic Fuels and Diamond Green Diesel, and internationally by Neste Oil. The KiOR technology, utilizing pyrolysis, is at an initial commercial scale. Other technology routes are not yet commercial and display a wide range of estimated costs in public sources. As these technology pathways mature and become more widespread, more specific information regarding their economics will be available, which would enable a more detailed analysis and performance comparison.

5) From a feedstock perspective, enough lignocellulosic material is projected to be available in support of the RFS mandate of 21 billion gallons of advanced biofuels. Crop and forest residues alone could yield about 8-24 billion gallons of biomass-based diesel/jet fuel in 2022 (assuming a conversion via fast pyrolysis). This potential could be larger if the conversion technologies achieve higher yields and if additional feedstock, such as dedicated energy crops, become available. However, there will be competition for lignocellulosic feedstock with the ethanol industry and renewable gasoline producers to meet the RFS mandate. Thus, it is unclear what share the renewable diesel/jet fuel would have in the total biofuels contribution. Ultimately, it will depend on the rate of commercialization of these technologies, selling price, and the transportation market demands.
6) Based on current statistics, and as proven by the biodiesel industry, there is enough lipid feedstock to support the production of 1 billion gallons of biomass-based diesel mandated by the RFS. Today, roughly half of the biodiesel in the United States is produced from soybeans. The remaining portion consists of animal fat, used cooking oil, canola, and some other minor feedstocks. While soybean production is projected to grow in coming years, the biodiesel industry hopes to achieve higher output through advanced technologies for increasing oil supply and production of new feedstock. If algal oil becomes commercially available, as projected within the next 5-10 years, it would greatly benefit both biodiesel and renewable diesel/jet fuel industries. Given the right resources, algal oil productivity can be quite high. Algae are a potential aquatic oil crop, but may also yield carbohydrates that can be converted to sugar.
7) Demand for diesel and jet fuel in the United States is projected to grow. As easily recoverable crude oil resources are diminishing and as their prices rise, more substitutes are expected to enter the market.
    - Among the diesel consumers in the country, freight trucking has the largest share. The number of light-duty vehicles using diesel is projected to increase, the rate of which will depend on the market penetration of other alternatives such as hybrid and electric vehicles.
    - Jet fuel is forming as a large and profitable market for the renewable fuels industry. Success in this area could stimulate a significant increase in the production of biofuels and associated feedstock. It is expected that jet fuel consumption by commercial

carriers will continue to grow over the next years, whereas jet fuel consumption by the military will remain flat.

8) For biomass-based diesel and jet fuel to be successful among the trucking and aviation companies, they must be cost-competitive with petroleum-based fuels. It is uncertain what the future holds for these substitutes, but it is expected that the next several years, as more facilities come online, will answer many questions about the economic viability of these technologies. Much will depend on the rate of recovery of U.S. and world economies, oil prices, carbon market, and political climate.

References

Advanced Biofuels & Biobased Materials Project Database. *Biofuels Digest.* (2012). Release Q3 2012: http://bit.ly/MP7MHv.

Agusdinata, D.; Zhao, F.; Lleleji, K. (2011). "Life Cycle Assessment of Potential Biojet Fuel Production in the United States." *Environmental Science and Technology* (45); pp. 9133-9143: http://pubs.acs.org/doi /abs/10.1021/es202148g.

American Trucking Association (ATA). (2012). "Renewable Diesel and Biodiesel." Accessed December 2012: http://www.trucking.org/advissues /energy

Amyris. "Breakthrough Science." Accessed July 2012, http://www.amyris biotech.com/Innovation/155/BreakthroughScience.

Anex, R. P.; Aden, A.; Kazi, F. K.; Fortman, J.; Swanson, R. M.; Wright, M. M.; Satrio, J. A.; Brown, R. C.; Daugaard, D. E.; Platon, A.; Kothandaraman, G.; Hsu, D. D.; Dutta, A. (2010). "Techno-Economic Comparison of Biomass-to-Transportation Fuels Via Pyrolysis, Gasification, and Biochemical Pathways." *Fuel.* 89 (1); pp. S29-S35: http://www.nrel.gov/docs/fy11osti/47764.pdf.

Biddy, M.; Jones, S. (2013). " Catalytic Upgrading of Sugars to Hydrocarbons Technology Pathway." Technical Report NREL/TP-5100-58055, PNNL-22319: http://www.nrel.gov/docs/fy13osti/58055.pdf.

Brown, T.R.; Wright, M.M.; Brown, R.C. (2011). "Estimating profitability of two biochar production scenarios: slow pyrolysis *vs* fast pyrolysis." *Biofuels, Bioproducts and Biorefining* (5); pp.54–68: http://onlinelibrary. wiley.com/doi/10.1002/bbb.254/pdf.

Cobalt. "BIOBUTANOL — Versatile Chemical Serving Large Markets". Accessed March 2013: http://www.cobalttech.com/.

Coskata Inc. "The Coskata Process." Accessed January 2011. http://coskata.com/process/.

Davis, R. (2009). "Techno-economic analysis of current technology for Fischer-Tropsch fuels production," NREL Technical Memorandum. National Bioenergy Center. August 14, 2009.

Davis, R.; Aden, A.; Pienkos, P. (2011). "Techno-economic analysis of autotrophic microalgae for fuel production," *Applied Energy* (88); pp. 3524–3531: http://www.sciencedirect.com/science

Davis, R.; Fishman, D.; Frank, E.; Wigmosta, M. (2012). "Renewable Diesel from Algal Lipids: An Integrated Baseline for Cost, Emissions, and Resource Potential from a Harmonized Model." Technical Report ANL/ESD/12-4, NREL/TP-5100-55431, PNNL21437, http://www.nrel.gov/docs/fy12osti/55431.pdf.

Defense Logistics Agency (DLA). (2012). "Fact Book Fiscal Year 2012". http://www.energy 2.pdf.

Dynamic Fuels. (2013). Dynamic Fuels, LLC website. Accessed November 2013. http://www.dynamicfuelsllc.com/about.aspx.

EIA. (2011). "Annual Energy Outlook 2011," April 2011, http://www.eia.gov/forecasts/aeo/.

EIA. (2013b). "Distillate Fuel Oil and Kerosene Sales by End Use," November 2013, http://www.eia.gov/dnav/pet/pet_cons_821use_dcu_nus_a.htm.

EIA. (2013c). "Product Supplied," September 2013, http://www.eia.gov/dnav/pet/pet_cons_psup_dc_nus_mbbl_a.htm.

EIA. (2013d). "State Energy Data System – Jet Fuel Consumption, Price, and Expenditure Estimates, 2012," September 2013, http://www.eia.gov/state fuel/html/fuel jf.html.

EIA. (2013e). "U.S. On-Highway Diesel Fuel Prices," October 2013, http://www.eia.gov/petroleum

Energy Information Administration (EIA). (2013a). "Monthly Energy Review – Biodiesel Overview," October 2013, http://www.eia.gov/totalenergy/data/monthly/#renewable.

Environmental Protection Agency (EPA). (2010a). "Regulation of Fuels and Fuel Additives: 2011 Renewable Fuel Standards," July 2010, http://www.gpo.gov/fdsys/pkg/FR-2010-07-20/pdf/2010-17281.pdf.

EPA (2009). "Renewable Fuel Standard Program (RFS2): Notice of Proposed Rulemaking", http://www.epa.gov/otaq/renewablefuels/rfs2 1-5.pdf.

EPA. (2010b). "Renewable Fuel Standard Program (RFS2)–2010 and Beyond," April 2010, presentation to the Union League Club, http://www.mcguirewoods.com/media. pdf.

EPA. (2013a). "RFS2 EMTS Informational Data," http://www.epa.gov/otaq /fuels/rfsdata/index.htm.

EPA. (2013b). "EPA Finalizes 2013 Renewable Fuels Standards," August 2013, http://www.epa.gov/otaq/fuels/renewablefuels/documents/420f1 3042.pdf.

EPA. (2013c). "EPA Proposes 2014 Renewable Fuels Standards, 2015 Biomass-Based Diesel Volume". November 2013. http://www.epa.gov /otaq/fuels/renewablefuels/documents/420f13048.pdf.

European Biofuels Technology Platform (EBTP). (2009). "Synthetic Biology, 'Modified Metabolism' and Plant Bioetchnology for Production of Advanced Biofuel Molecules and Improved Feedstocks,"Accessed December 2012, http://www.biofuelstp.eu/biotechnology

FAPRI – Iowa State University. (2012). "2012 World Agricultural Outlook," http://www.fapri.iastate.edu/outlook/2012/.

Food and Agricultural Policy Research Institute at the University of Missouri-Columbia (FAPRI-MU). (2009). "Renewable Identification Number Markets: Draft Baseline Table," http://www.fapri.missouri.edu/outreach f.

Gevo. (2011). "Renewable Solutions." http://gevo.com/wpcontent/uploads /2011/05/GEVO-wp-iso-ftf.pdf

Greentech Media. (2010). "LS9's Genetic Breakthrough: Will it Produce Biofuels as Scale?" http://www.greentechmedia.com/articles/read/LS9s-Biofuel-ProductionBreakthrough/.

Hart Energy Consulting. (2009). "Global Biofuels Outlook 2010-2020", http://www.hartenergy.com/Downstream/Research-And-Consulting/ Global-BiofuelsOutlook-2010-2020/.

"How the New MPG Standards Will Affect Drivers, Automakers, Car Dealerships & More." *Time Magazine*. (2012). August 30 2012, http://business.time.com/2012/08/30/how-the-new-mpg-standards-will-affect-driversautomakers-car-dealerships-more/.

Jegannathan, K.; Eng-Seng, C.; Ravindra, P. (2011) "Economic assessment of biodiesel production: Comparison of alkali and biocatalyst processes." *Renewable and Sustainable Energy Review,* (15); pp. 745–751. http://www.sciencedirect.com/science

Journal of Commerce (JOC). (2013). "Top 50 Trucking Companies". Accessed September 2013: http://www.joc.com/trucking-logistics/

Kalnes, T.; Marker, T.; Shonnard, D.R. (2007). Article A48, *International Journal of Chemical Reactor Engineering,* (5).

KiOR Inc. "Production Facilities". Accessed March 2013. http://www.kior.com/content/?s=6&s2=56&p=56&t=Production-Facilities

Marker, T.; Felix, L.; Linck, M.; Ortaz Toral, P.; Wangerow, J.; Kraus, L.; McLeod, C.; DelPaggio, A.; Tan, E.; Gephart, J.; Gromov, D.; Roberts, M., Purtle, I., Starr, J., Hahn, J., Dorrington, P., Stevens, J., Shonnard, D., and Maleche, E. (2012) "Biomass to Gasoline and Diesel Using Integrated Hydropyrolysis and Hydroconversion." Extended Abstract, AICHE. Gas Technology Institute (GTI) Technical Report, DOI 10.2172/1059031. http://www.osti.gov/bridge/product.biblio.jsp?osti_id=1059031

Milbrandt, A. (2005). "Geographic Perspective on the Current Biomass Resource Availability in the United States." Golden, CO: National Renewable Energy Laboratory (NREL). NREL/TP-560-39181; 70 pp: http://www.nrel.gov/docs/fy06osti/39181.pdf.

NABC. (2012b). "Hydrothermal Liquefaction – Results from Stage I Extension". January 5, 2012. http://www.nabcprojects.org/pdfs /hydrothermal_liquefaction_results_stage_i_extension.p df

NAS. (2009). "Liquid transportation fuels from coal and biomass: technological status, costs, and environmental impacts." Washington DC, 2009 http://sites.nationalacademies.org/xpedio/groups/energysite/documents/webpage/energy 054519.pdf

National Advanced Biofuels Consortium (NABC). (2012a). "A Refiner's Perspective on Advanced Biofuels", Prepared from a talk delivered at the at the January 2012 annual NABC meeting by Rick Weyen of Tesoro. http://www.nabcprojects.org/pdfs/refiner_perspective_advanced_biofuels.pdf

National Energy Technology Laboratory (NETL). (2009). "Affordable, Low-Carbon Diesel Fuel from Domestic Coal and Biomass". DOE/NETL-2009/1349. http://www.netl.doe.gov/energy

Neste Oil. "NExBTL Aviation Fuel". Accessed January 2012, http://www.nesteoil.com/default.asp?path=1,41,11991,12243,17555

Pacific Northwest National Laboratory (PNNL). (2009). "Production of Gasoline and Diesel from Biomass via Fast Pyrolysis, Hydrotreating and Hydrocracking: A Design Case". DOE/PNNL-18284. http://www.pnl.gov /main/publications/external/technical_reports/PNNL-18284.pdf

Pearlson,M., Wollersheim, C., and Hileman, J. (2012) "A techno-economic review of hydroprocessed renewables esters and fatty acids for jet fuel production." Biofuels, Bioproducts, and Biorefineries 7:89-96 (2013); DOI: 10.1002/bbb. http://onlinelibrary.wiley.com/doi/10.1002 /bbb.1378 /pdf

PR Newswire. (2012). "U.S. Clean Diesel Auto Sales Increase 25.6 Percent in 2012", http://www.prnewswire.com/news-releases/us-clean-diesel-auto-sales-increase-256- percent-in-2012-181048881.html

Seeking Alpha. (2012). "The U.S. EPA Approves Solazyme's Fuel For Commercial Use." http://seekingalpha.com/article/778491-the-u-s-epa-approves-solazymes-fuel-forcommercial-use

Smagala, T.G.; Christensen, E.; Christison, K.M.; Mohler, R.E.; Gjersing, E.; McCormick, R.L. (2013). "Hydrocarbon Renewable and Synthetic Diesel Fuel Blendstocks: Composition and Properties." *Energy Fuels* 27 (1); pp. 237-246.

Soybean Review. (2011). "RFS2 Provides Stability for Biodiesel's Future", http://soybeanreview.com/article/rfs2-provides-stability

Syntroleum. (2011). "Syntroleum Announces Record July Production and Second Quarter Results," http://www.syntroleum.com/

Syntroleum. (2013). Syntroleum Corporation 2013 Form 10-K (Annual Report). Filed 03/15/13 for the Period Ending 12/31/12.

Tao, L.; Aden, A. (2009). "The economics of current and future biofuels." *In Vitro Cellular & Developmental Biology – Plant*, (45); pp. 199–217. http://link.springer.com/article/10.1007%2Fs11627-009-9216-8

"Terrabon achieves production milestone at Texas demo facility." *Ethanol Producer Magazine*. (2011). January 27, 2011. http://www.ethanol

The City Wire. (2013a). "Dynamic Fuels Plant on Standby, Indefinitely." August 7, 2013, http://www.thecitywire.com/node/28947#.Upzkn8SkqDs.

The City Wire. (2013b). "Tyson's Dynamic Fuels Plant Burns Cash." November 18, 2013, http://www.thecitywire.com/node/30545 #.Upzkic SkqDs.

The National Academy of Sciences (NAS). (2011). Committee on Economic and Environmental Impacts of Increasing Biofuels Production; National Research Council. "Renewable Fuel Standard: Potential Economic and Environmental Effects of U.S. Biofuel Policy," The National Academies Press, pp. 131-134, http://www.nap.edu/catalog.php?record_id=13105

U.S. Census Bureau. (2010). "M311K - Fats and Oils: Production, Consumption, and Stocks," http://www.census.gov/manufacturing

U.S. Department of Agriculture (USDA). (2012). "Oil Crops Yearbook 2011" March 2012, http://usda.mannlib.cornell.edu/MannUsda/viewDocument

Info.do;jsessionid=53C6FDE58607C6FDCDBF42000CF26338?documentID=1290.

U.S. Department of Energy (USDOE). (2010). "Diesel Power: Clean Vehicles for Tomorrow," http://www1.eere.energy

USDA. (2013). Economic Research Service. "Oil Crops Yearbook 2012". March 2013, http://www.ers.usda.gov/data-products/oil RDsqg

USDOE. (2011). *U.S. Billion-Ton Update: Biomass Supply for a Bioenergy and Bioproducts Industry*. R.D. Perlack and B.J. Stokes (Leads), ORNL/TM-2011/224. Oak Ridge National Laboratory, Oak Ridge, TN, http://www1.eere.energy

Wigmosta, M.; Coleman, A.; Skaggs, R.; Huessemann, M.; Lane, L. (2011). "National Microalgae Biofuel Production Potential and Resource Demand", *Water Resources Research* (47): http://www.agu.org /journals/wr/wr1104/2010WR009966/2010WR009966.pdf

Wright, M.; Brown, R.; Boateng, A. (2008). "Distributed processing of biomass to bio-oil for subsequent production of Fischer-Tropsch liquids."*Biofuels, Bioproducts, and Biorefining* (2); pp. 229–238. http:// onlinelibrary.wiley.com/doi/10.1002/bbb.73/pdf

Wright, M.; Daugaard, D.; Satrio, J.; Brown, R. (2009). "Techno-economic analysis of biomass fast pyrolysis to transportation fuels." Proceedings of the 238th National Meeting and Exposition of the American Chemical Society.

Wright, M.; Daugaard, D.; Satrio, J.; Brown, R. (2010). "Techno-economic analysis of biomass fast pyrolysis to transportation fuels". *Fuel* (89); pp. S2–S10. http://www.sciencedirect.com/science

Yusuf, N.; Kamarudin, S.; Yaakub, Z. (2011). "Overview on the current trends in biodiesel production." *Energy Conversion and Management,* (52); pp. 2741–2751.

ZeaChem. (2011). "Technology Overview." Accessed January 2011, http://www.zeachem.com/technology

APPENDIX

Table 8. Distillate Fuel Oil and Kerosene Sales by End Use

	2007	2008	2009	2010	2011	2012
Residential						
Distillate Fuel Oil	5,141,842	5,568,664	4,183,891	3,930,517	3,625,747	3,473,310
No. 1	81,992	83,379	83,828	65,325	67,530	53,219
No. 2	5,059,651	5,484,687	4,020,053	3,865,192	3,558,216	3,420,091
Kerosene	326,320	157,606	206,136	216,887	137,232	67,316
Commercial						
Distillate Fuel Oil	2,718,874	2,859,295	2,785,246	2,738,384	2,715,335	2,557,543
No. 1 Distillate	64,080	86,322	93,323	70,671	63,082	60,566
No. 2 Distillate	2,475,571	2,530,529	2,559,629	2,545,295	2,524,697	2,384,977
No. 2 Fuel Oil	1,217,835	1,137,514	890,809	885,331	771,343	621,683
Ultra Low Sulfur Diesel	731,682	1,029,830	1,311,244	1,418,616	1,630,060	1,662,447
Low Sulfur Diesel	245,223	319,052	207,287	175,838	85,497	83,036
High Sulfur Diesel	280,932	162,132	130,288	65,520	35,958	17,881
No. 4 Fuel Oil	179,014	149,644	142,295	122,420	126,655	111,970
Residual Fuel Oil	481,368	403,977	415,197	366,363	316,713	296,150
Kerosene	67,969	32,857	31,291	20,716	23,848	8,988
Industrial						
Distillate Fuel Oil	2,464,904	2,593,759	2,159,439	2,046,164	2,179,953	2,325,583
No. 1 Distillate	37,787	35,510	26,831	32,532	63,588	39,364
No. 2 Distillate	2,371,370	2,534,715	2,054,033	1,948,268	2,050,831	2,215,033
No. 2 Fuel Oil	324,891	310,685	154,760	132,870	143,117	125,447
Low Sulfur Diesel	1,318,312	1,833,331	1,663,665	1,671,967	1,849,222	2,047,177
High Sulfur Diesel	727,167	390,699	266,278	153,432	63,492	42,410
No. 4 Fuel Oil	67,749	23,826	77,844	64,363	60,634	71,186
Residual Fuel Oil	1,187,319	1,043,883	726,210	667,672	772,676	484,957
Kerosene	88,372	22,445	25,536	45,145	22,587	12,820
Farm						
Distillate Fuel Oil	3,202,847	3,744,636	2,660,024	2,928,175	2,942,436	3,031,878
Diesel	3,147,431	3,698,265	2,620,378	2,895,184	2,912,647	3,007,499
Other Distillate	55,416	46,471	39,647	33,071	29,789	24,379
Kerosene	8,531	4,893	6,414	6,763	3,410	1,712
Electric Power						
Distillate Fuel Oil	669,061	816,626	681,386	648,144	606,683	461,684
Residual Fuel Oil	2,746,574	1,564,584	1,111,364	1,030,687	671,585	467,776
Oil Company						
Distillate Fuel Oil	774,904	1,086,809	760,077	951,322	1,381,127	1,710,513
Residual Fuel Oil	43,977	57,614	26,168	20,783	19,759	17,031
Total Transportation (Railroad, Vessel Bunkering, On-Highway)						
Distillate Fuel Oil	45,209,234	42,741,511	38,818,930	40,560,181	41,414,854	41,229,545
Residual Fuel Oil	6,326,931	5,257,610	4,589,049	5,142,573	4,560,070	4,819,560
Railroad						
Distillate Fuel Oil	3,034,512	3,229,625	2,758,140	2,974,841	3,121,150	3,118,150
Vessel Bunkering						
Distillate Fuel Oil	1,573,981	1,983,427	1,912,984	2,002,834	2,133,396	1,768,324
Residual Fuel Oil	6,326,931	5,257,610	4,589,049	5,142,573	4,560,070	4,819,560
On-Highway						
Distillate Fuel Oil	39,601,744	37,528,464	34,147,806	35,582,625	36,160,308	36,343,072
Military						
Distillate Fuel Oil	383,145	278,575	243,728	243,242	246,243	142,695
Diesel	202,873	264,137	213,643	215,218	223,841	126,268
Other Distillate	180,272	15,438	30,085	28,025	22,403	16,429
Residual Fuel Oil	17,719	8,250	14,889	9,851	14,653	10,324
Off-Highway						
Distillate Fuel Oil	2,512,394	2,605,669	1,985,592	2,148,677	2,070,260	2,088,157
Construction	2,206,899	2,244,505	1,749,599	1,809,806	1,810,394	1,852,241
Non-Construction	306,406	367,166	236,093	338,870	269,866	236,016
All Other						
Distillate Fuel Oil	0	0	0	0	0	0
Residual Fuel Oil	2,606	3,749	6,563	6,860	2,664	1,418
Kerosene	1,520	1,036	633	2,297	899	245

In thousand gallons. Data source: EIA 2013b.

Table 9. No. 2 Diesel Sales for On-Highway Use

	2007	2008	2009	2010	2011	2012
U.S.	39,801,744	37 528,464	34,147,806	35,582,625	36,160,308	36,343,072
East Coast (PADD 1)	11,739,455	10 763,333	9,929,426	10,367,337	10,332,863	10,257,620
New England (PADD 1A)	1,123,563	1 062,422	1,044,171	1,052,933	1,064,679	1,063,943
Connecticut	303,570	292,688	266,121	267,948	271,070	266,474
Maine	181,010	180,284	176,004	178,774	176,241	173,717
Massachusetts	402,629	375,414	399,575	394,967	410,375	418,613
New Hampshire	97,695	97,837	92,882	93,381	91,778	91,143
Rhode Island	73,882	55,826	55,684	59,220	57,035	54,666
Vermont	64,777	60,373	53,905	58,643	58,180	59,330
Central Atlantic (PADD 1B)	4,232,345	3 956,845	3,629,109	3,796,971	3,883,499	3,898,410
Delaware	66,271	59,569	55,610	55,342	55,904	56,530
District of Columbia	10,710	15,887	11,944	11,975	8,742	8,535
Maryland	573,321	528,862	515,315	536,150	501,897	503,999
New Jersey	988,732	902,201	679,361	793,752	888,154	784,867
New York	1,118,505	1 120,424	1,064,997	1,083,536	1,082,247	1,100,897
Pennsylvania	1,474,806	1 329,902	1,301,882	1,316,216	1,346,555	1,443,582
Lower Atlantic (PADD 1C)	6,383,547	5 744,066	5,256,146	5,517,433	5,384,685	5,295,267
Florida	1,670,536	1 478,514	1,322,703	1,340,494	1,329,312	1,340,337
Georgia	1,508,118	1 325,065	1,197,220	1,260,672	1,187,538	1,121,051
North Carolina	1,076,721	968,520	886,949	940,838	954,190	920,147
South Carolina	722,868	667,949	641,944	715,795	715,136	647,447
Virginia	1,111,050	1 017,285	935,552	969,057	917,431	971,673
West Virginia	294,254	286,733	271,778	290,577	281,078	294,612
Midwest (PADD 2)	12,627,117	12 123,237	10,905,027	11,509,060	11,784,798	11,886,665
Illinois	1,516,293	1 436,180	1,364,571	1,340,323	1,454,548	1,387,514
Indiana	1,356,498	1 302,135	1,109,103	1,207,711	1,225,199	1,247,331
Iowa	648,924	643,923	601,324	633,299	641,795	643,505
Kansas	489,575	480,947	441,256	476,191	467,779	474,517
Kentucky	875,135	821,452	742,976	785,240	772,677	767,231
Michigan	906,538	838,960	780,954	819,342	820,526	837,612
Minnesota	673,758	663,548	584,807	610,447	639,520	653,051
Missouri	1,087,199	985,190	944,991	977,111	970,936	957,624
Nebraska	434,093	406,238	386,874	425,442	421,847	418,302
North Dakota	177,467	193,615	194,777	237,443	318,231	376,496
Ohio	1,597,741	1 494,384	1,323,991	1,437,404	1,455,547	1,448,059
Oklahoma	840,366	906,265	729,736	741,369	789,449	814,298
South Dakota	202,607	198,713	197,232	213,298	207,311	227,663
Tennessee	1,061,941	1 002,320	826,077	891,637	895,097	882,475
Wisconsin	758,982	749,367	676,358	712,803	704,336	750,987
Gulf Coast (PADD 3)	7,615,389	7 338,939	6,641,445	6,947,707	7,234,843	7,428,440
Alabama	848,402	736,640	657,070	711,371	717,466	705,904
Arkansas	673,079	627,835	592,469	618,731	605,275	605,758
Louisiana	703,016	690,551	694,073	732,017	749,400	667,605
Mississippi	625,746	623,455	553,866	562,673	552,909	578,156
New Mexico	531,013	473,827	432,794	472,924	495,600	495,026
Texas	4,234,133	4 186,631	3,711,173	3,849,991	4,114,193	4,375,991
Rocky Mountain (PADD 4)	1,999,274	1 891,875	1,693,227	1,768,306	1,800,630	1,854,474
Colorado	588,910	566,484	510,211	523,056	501,361	517,545
Idaho	275,706	243,868	225,624	254,417	254,007	255,815
Montana	263,696	250,069	235,067	243,660	251,686	259,418
Utah	480,993	446,071	399,528	401,810	474,895	462,240
Wyoming	389,969	385,383	322,797	345,363	318,681	359,456
West Coast (PADD 5)	5,820,509	5 411,080	4,978,681	4,990,215	5,007,174	4,915,873
Alaska	174,112	171,630	202,102	166,599	169,158	111,113
Arizona	865,536	792,698	739,863	741,568	757,789	742,679
California	3,091,491	2 838,723	2,591,988	2,602,646	2,633,352	2,603,546
Hawaii	52,692	56,394	46,847	50,187	45,792	47,897
Nevada	378,182	329,292	302,145	298,895	288,458	289,875
Oregon	560,598	534,041	498,335	513,521	513,970	507,784
Washington	697,898	688,302	597,401	616,779	598,655	612,979

In thousand gallons. Data source: EIA 2013b.

Table 10. Diesel Sales for Farm Use

	2007	2008	2009	2010	2011	2012
U.S.	3,147,431	3 698,265	2,620,378	2,895,104	2,912,647	3,007,499
East Coast (PADD 1)	358,009	377,605	326,068	449,575	368,704	374,504
New England (PADD 1A)	18,818	20,214	11,470	10,956	14,953	13,938
Connecticut	1,798	2,160	987	1,091	1,515	1,880
Maine	6,509	7,547	2,350	3,713	4,335	4,399
Massachusetts	3,273	1,240	1,223	831	1,195	1,831
New Hampshire	2,229	2,622	2,083	772	1,263	1,295
Rhode Island	40	103	20	16	23	34
Vermont	4,970	6,543	4,808	4,532	6,622	4,499
Central Atlantic (PADD 1B)	96,794	113,257	90,495	99,578	104,693	99,395
Delaware	5,839	4,762	5,903	6,821	8,541	6,767
District of Columbia	--	0	0	0	0	0
Maryland	13,677	17,618	8,020	10,720	11,420	11,292
New Jersey	2,030	2,726	5,951	7,343	6,115	6,413
New York	39,923	43,886	30,246	20,092	29,844	30,636
Pennsylvania	35,325	44,266	40,376	54,603	48,773	44,287
Lower Atlantic (PADD 1C)	242,397	244,133	224,103	339,041	249,059	261,171
Florida	69,001	87,131	86,642	204,791	109,128	103,325
Georgia	68,172	47,824	69,053	62,281	63,725	79,470
North Carolina	49,930	50,645	41,443	39,454	38,571	42,955
South Carolina	20,374	20,058	8,176	9,379	12,217	10,832
Virginia	33,347	36,810	17,192	20,263	23,370	22,452
West Virginia	1,573	1,667	1,596	2,874	2,048	2,137
Midwest (PADD 2)	1,596,524	1 844,300	1,384,186	1,502,450	1,510,797	1,599,681
Illinois	175,985	210,548	119,015	142,946	152,491	151,626
Indiana	136,331	130,194	92,486	85,257	116,498	126,061
Iowa	155,346	201,338	178,089	200,458	191,678	202,588
Kansas	148,827	189,901	138,644	156,717	135,899	135,168
Kentucky	24,308	30,569	20,187	18,594	20,430	18,582
Michigan	47,329	53,828	49,200	56,779	60,122	50,302
Minnesota	114,924	131,577	153,716	166,789	158,840	157,524
Missouri	138,517	123,377	88,445	91,575	92,162	101,940
Nebraska	199,851	233,656	149,308	142,888	142,327	199,787
North Dakota	100,554	136,067	96,106	119,470	134,935	126,979
Ohio	107,688	130,449	95,777	106,136	87,699	106,475
Oklahoma	57,403	56,885	36,066	43,851	26,505	33,172
South Dakota	59,635	67,897	63,165	56,899	67,198	64,939
Tennessee	36,162	31,532	20,400	23,276	21,851	23,379
Wisconsin	93,464	116,481	83,583	90,816	102,163	101,158
Gulf Coast (PADD 3)	604,338	735,740	381,868	434,193	462,577	468,982
Alabama	32,659	44,138	17,882	19,881	24,518	24,503
Arkansas	212,255	222,990	68,641	92,965	83,323	87,118
Louisiana	51,507	58,838	41,219	42,255	48,056	45,753
Mississippi	50,123	41,385	41,080	57,087	52,559	81,878
New Mexico	15,472	20,028	11,277	14,821	10,950	12,816
Texas	242,322	348,361	201,769	207,183	243,170	216,915
Rocky Mountain (PADD 4)	207,302	227,081	203,467	167,704	175,686	159,171
Colorado	43,695	58,955	40,518	39,603	41,532	32,759
Idaho	54,577	55,455	53,759	66,442	70,724	61,287
Montana	94,618	96,116	93,447	44,921	44,840	47,207
Utah	5,885	5,110	6,786	5,831	6,794	7,031
Wyoming	8,526	11,445	8,957	10,907	11,797	10,886
West Coast (PADD 5)	381,259	513,539	324,788	341,182	394,882	405,161
Alaska	37	65	108	109	126	177
Arizona	19,184	35,197	35,893	34,514	39,760	35,964
California	268,177	349,107	187,956	205,254	244,358	258,004
Hawaii	4,102	3,570	4,175	3,493	3,483	3,074
Nevada	3,008	2,901	5,322	6,351	6,444	5,102
Oregon	26,982	30,579	33,964	33,174	46,209	47,498
Washington	59,769	92,119	57,371	58,287	54,502	55,342

In thousand gallons. Data source: EIA 2013b.

Table 11. Sales of Distillate Fuel Oil for Railroad Use

	2007	2008	2009	2010	2011	2012
U.S.	3,634,512	3,229,625	2,759,140	2,974,641	3,121,150	3,118,150
East Coast (PADD 1)	580,632	500,071	459,324	482,929	514,418	492,156
New England (PADD 1A)	69,282	47,582	43,763	53,930	51,126	33,306
Connecticut	4,450	3,219	2,219	2,006	2,006	5,195
Maine	126	1,694	7,252	8,284	6,818	5,970
Massachusetts	63,896	40,378	24,852	33,130	32,647	12,307
New Hampshire	119	126	697	86	124	116
Rhode Island	13	72	4	24	3	133
Vermont	678	2,092	8,740	10,400	9,528	9,586
Central Atlantic (PADD 1B)	210,461	177,750	152,309	196,570	233,005	204,527
Delaware	1,404	1,120	1,096	879	126	149
District of Columbia	0	0	0	1,229	6,392	6,770
Maryland	11,546	3,214	17,035	34,717	36,283	20,384
New Jersey	15,616	15,055	8,071	1,778	1,660	1,325
New York	63,226	44,510	35,307	33,709	42,254	35,237
Pennsylvania	118,670	113,851	90,800	124,258	146,291	140,663
Lower Atlantic (PADD 1C)	300,889	274,739	263,252	232,429	230,287	254,322
Florida	74,409	64,963	33,651	42,353	46,461	66,711
Georgia	78,927	69,710	62,072	63,770	71,374	63,902
North Carolina	47,855	29,022	89,823	62,103	32,158	41,501
South Carolina	11,321	16,023	3,602	3,051	3,973	3,983
Virginia	72,611	79,606	63,960	49,503	63,611	67,769
West Virginia	15,766	15,416	10,143	11,650	12,711	10,456
Midwest (PADD 2)	1,561,277	1,420,396	1,144,926	1,223,206	1,215,528	1,195,263
Illinois	40,116	51,287	55,322	72,188	58,526	63,808
Indiana	65,820	51,232	37,773	50,736	63,437	68,061
Iowa	58,640	62,458	40,494	41,663	36,136	30,156
Kansas	92,323	129,141	147,106	78,143	80,404	99,475
Kentucky	170,042	94,124	48,002	42,101	67,347	61,840
Michigan	49,528	41,887	25,920	18,376	10,330	13,352
Minnesota	123,390	78,651	39,188	47,567	61,340	92,275
Missouri	27,467	13,281	19,765	36,396	51,179	44,914
Nebraska	12,732	27,507	75,064	214,176	181,421	166,060
North Dakota	124,832	58,667	12,849	8,983	9,839	43,907
Ohio	333,069	316,926	206,134	179,048	203,135	175,258
Oklahoma	348,832	395,252	352,301	349,077	313,806	245,376
South Dakota	8,572	10,024	5,730	5,860	7,182	10,826
Tennessee	76,692	41,676	53,391	54,332	58,125	52,522
Wisconsin	29,222	48,285	25,886	24,559	13,321	27,434
Gulf Coast (PADD 3)	699,882	631,796	542,036	573,037	694,053	729,109
Alabama	59,852	42,588	44,546	42,465	97,177	125,439
Arkansas	20,237	27,693	25,148	18,302	26,907	43,494
Louisiana	43,862	32,201	18,345	25,425	32,515	28,110
Mississippi	46,730	31,617	24,727	17,936	37,741	29,848
New Mexico	6,152	2,092	245	1,780	1,707	19,123
Texas	523,049	495,604	429,026	467,128	498,006	483,096
Rocky Mountain (PADD 4)	262,644	222,054	212,571	228,200	245,446	214,160
Colorado	4,014	5,422	47,830	66,510	71,365	77,038
Idaho	21,070	14,622	9,678	31,307	30,448	25,068
Montana	107,710	94,818	68,520	58,543	65,919	41,901
Utah	44,721	24,643	21,178	24,774	33,371	24,216
Wyoming	85,129	82,549	65,365	47,065	44,344	45,938
West Coast (PADD 5)	530,077	455,308	400,283	467,270	451,704	487,461
Alaska	6,419	6,120	5,899	5,399	5,754	5,564
Arizona	11,940	7,230	8,200	10,566	9,698	20,624
California	317,292	261,225	219,854	252,057	255,313	258,354
Hawaii	4	6	5	37	4	4
Nevada	6,874	7,101	11,594	7,446	8	44
Oregon	82,369	94,925	91,305	104,445	87,470	114,507
Washington	105,180	78,701	63,425	87,321	93,458	88,364

In thousand gallons. Data source: EIA 2013b.

Table 12. Sales of Distillate Fuel Oil for Off-Highway Use

	2007	2008	2009	2010	2011	2012
U.S.	2,512,394	2,605,660	1,985,592	2,148,677	2,070,260	2,088,157
East Coast (PADD 1)	833,519	883,356	605,884	615,812	634,470	621,261
New England (PADD 1A)	92,754	113,790	81,453	102,263	102,751	75,212
Connecticut	21,159	19,948	14,456	16,124	16,435	10,683
Maine	12,193	15,262	14,483	15,495	16,622	18,373
Massachusetts	39,016	56,006	27,388	43,133	43,432	19,129
New Hampshire	11,495	14,814	8,898	12,689	11,421	10,558
Rhode Island	4,540	2,129	5,652	3,821	3,725	3,057
Vermont	4,352	5,632	10,576	11,000	11,116	13,413
Central Atlantic (PADD 1B)	226,685	252,027	186,785	187,163	213,795	208,407
Delaware	3,149	3,210	2,578	2,201	2,306	1,812
District of Columbia	1,988	1,223	1,043	357	920	892
Maryland	35,368	37,065	15,198	14,402	23,521	21,838
New Jersey	65,404	65,491	48,769	47,895	62,894	59,809
New York	51,681	48,869	37,033	34,272	38,054	35,089
Pennsylvania	69,096	96,169	82,164	88,035	86,101	88,967
Lower Atlantic (PADD 1C)	514,080	517,539	337,646	326,386	317,924	337,641
Florida	124,123	135,397	112,263	110,675	104,005	109,119
Georgia	124,009	111,648	75,856	79,376	76,496	81,370
North Carolina	53,366	52,272	43,755	53,910	51,483	42,119
South Carolina	51,781	57,191	36,537	35,998	29,800	41,179
Virginia	138,601	139,537	59,539	38,195	46,783	56,139
West Virginia	22,201	21,494	9,695	8,231	9,357	7,715
Midwest (PADD 2)	686,931	716,068	544,191	557,179	524,627	554,255
Illinois	110,067	108,788	60,997	59,541	56,167	66,746
Indiana	33,460	50,166	52,110	41,855	49,234	49,992
Iowa	24,765	28,979	29,836	30,989	26,475	30,634
Kansas	21,572	17,213	28,999	30,359	25,422	19,184
Kentucky	26,757	34,351	31,866	27,906	25,590	21,723
Michigan	47,782	52,039	35,267	34,737	35,147	33,514
Minnesota	57,964	86,484	45,198	54,211	62,268	55,294
Missouri	68,262	64,660	50,840	61,485	41,060	31,435
Nebraska	44,362	11,846	26,223	18,344	14,586	19,173
North Dakota	11,741	11,056	25,623	27,802	25,811	44,192
Ohio	68,156	74,876	56,113	56,694	49,976	60,223
Oklahoma	44,661	45,228	25,704	24,332	24,912	27,796
South Dakota	10,282	9,260	9,622	7,239	15,484	10,211
Tennessee	65,938	68,110	31,117	43,251	37,539	43,024
Wisconsin	51,163	53,012	34,676	38,436	34,954	41,115
Gulf Coast (PADD 3)	480,332	486,148	360,795	382,691	370,941	417,101
Alabama	62,743	80,762	61,501	58,051	58,172	69,907
Arkansas	48,044	30,752	23,210	29,752	22,773	22,622
Louisiana	76,131	82,523	48,510	68,889	44,130	42,719
Mississippi	59,253	53,833	17,862	18,404	20,889	26,199
New Mexico	3,235	16,705	5,729	24,907	24,865	29,454
Texas	230,926	221,572	203,983	182,689	200,112	226,200
Rocky Mountain (PADD 4)	149,317	138,282	139,107	121,101	125,195	113,299
Colorado	48,550	47,759	52,113	48,837	56,745	45,876
Idaho	23,362	18,349	14,912	11,780	11,916	10,317
Montana	22,325	28,642	20,985	19,520	14,700	14,814
Utah	35,219	29,871	26,922	19,952	26,754	20,749
Wyoming	19,860	13,661	24,176	21,012	15,080	21,543
West Coast (PADD 5)	362,295	381,807	335,615	471,893	415,027	382,242
Alaska	16,599	22,832	14,334	14,738	10,126	13,962
Arizona	66,808	76,168	49,894	60,142	65,414	66,025
California	135,327	127,748	146,296	195,299	208,632	183,553
Hawaii	9,414	8,570	9,806	7,839	7,920	9,578
Nevada	29,733	29,091	43,385	124,123	44,879	37,652
Oregon	22,925	33,848	34,306	30,108	34,562	36,028
Washington	81,488	83,550	37,593	39,644	43,493	35,443

In thousand gallons. Data source: EIA 2013b.

Table 13. Distillate Fuel Oil Sales for Vessel Bunkering Use

	2007	2008	2009	2010	2011	2012
U.S.	1,923,981	1,983,422	1,912,984	2,002,834	2,133,395	1,768,324
East Coast (PADD 1)	466,132	461,533	276,013	259,319	296,947	283,254
New England (PADD 1A)	43,014	69,102	45,147	30,589	32,414	38,891
Connecticut	6,654	5,683	3,914	1,898	1,502	2,838
Maine	8,298	6,815	15,611	4,207	4,128	13,349
Massachusetts	21,336	48,094	19,193	17,529	17,132	13,612
New Hampshire	2,740	2,552	2,327	1,110	1,395	1,815
Rhode Island	3,987	5,958	4,101	5,824	8,257	7,243
Vermont	0	0	0	21	0	35
Central Atlantic (PADD 1B)	147,629	129,789	104,487	67,726	76,446	74,154
Delaware	615	919	582	485	1,658	615
District of Columbia	11	7	5	13	15	17
Maryland	21,380	16,507	8,240	8,335	2,662	7,300
New Jersey	87,549	74,725	68,321	37,276	61,990	55,842
New York	12,339	10,814	8,497	6,869	4,453	7,089
Pennsylvania	25,735	26,816	18,842	14,748	5,669	3,290
Lower Atlantic (PADD 1C)	275,489	262,642	126,379	161,005	188,087	170,209
Florida	145,269	180,514	84,718	118,991	142,198	131,685
Georgia	14,016	10,831	10,765	12,904	12,387	11,300
North Carolina	8,009	8,280	5,957	8,231	5,613	4,177
South Carolina	18,795	15,892	8,094	8,043	5,865	4,966
Virginia	43,971	18,558	16,745	12,731	21,562	18,063
West Virginia	45,429	28,568	99	105	461	18
Midwest (PADD 2)	386,653	428,712	385,719	370,587	389,493	354,341
Illinois	71,805	101,851	85,117	56,575	56,000	59,508
Indiana	29,805	24,226	17,165	19,849	22,319	8,139
Iowa	4,867	4,151	4,523	5,381	5,163	5,609
Kansas	0	0	0	0	0	0
Kentucky	91,516	104,387	102,305	98,294	104,160	96,669
Michigan	8,900	7,585	7,585	17,875	19,825	20,792
Minnesota	7,551	18,987	9,811	7,638	4,111	1,649
Missouri	34,158	42,760	44,967	49,922	42,388	40,858
Nebraska	0	0	0	0	0	0
North Dakota	0	0	0	0	0	0
Ohio	12,122	17,733	24,586	27,668	30,684	23,532
Oklahoma	10,764	10	8	9	6	7
South Dakota	0	0	0	0	0	0
Tennessee	114,703	106,326	88,312	85,786	103,077	95,856
Wisconsin	461	696	1,339	1,591	1,760	1,723
Gulf Coast (PADD 3)	613,864	721,875	827,977	1,010,776	1,026,408	677,343
Alabama	36,334	63,100	61,852	65,017	41,339	25,542
Arkansas	420	49,384	47,557	47,769	55,246	425
Louisiana	351,726	399,347	378,642	481,453	564,650	382,462
Mississippi	93,581	106,799	141,302	93,384	58,285	58,505
New Mexico	0	0	0	0	0	0
Texas	131,803	103,245	198,625	323,153	306,887	210,408
Rocky Mountain (PADD 4)	27	26	19	6	2	17
Colorado	2	2	1	1	2	2
Idaho	25	25	18	5	0	15
Montana	0	0	0	0	0	0
Utah	0	0	0	0	0	0
Wyoming	0	0	0	0	0	0
West Coast (PADD 5)	457,306	371,275	423,256	362,147	420,545	453,369
Alaska	127,614	119,460	104,126	115,129	118,900	136,681
Arizona	1	1	0	1	2	1
California	104,583	88,088	163,438	96,015	115,772	127,038
Hawaii	129,743	58,786	71,496	63,757	88,652	86,663
Nevada	25	22	7	6	6	7
Oregon	20,502	24,003	24,143	22,626	21,728	22,450
Washington	74,838	80,915	60,046	64,613	75,485	80,528

In thousand gallons. Data source: EIA 2013b.

Table 14. Diesel Sales for Military Use

	2007	2008	2009	2010	2011	2012
U.S.	202,873	254,537	213,643	215,218	223,841	126,268
East Coast (PADD 1)	43,728	57,003	47,877	42,842	58,296	35,470
New England (PADD 1A)	8,260	15,028	3,342	3,427	5,656	2,836
Connecticut	1,660	997	207	236	622	501
Maine	3,551	6,915	1,094	2,495	1,300	774
Massachusetts	2,125	3,182	500	343	3,101	271
New Hampshire	815	3,178	1,225	292	143	83
Rhode Island	19	399	35	21	468	1,091
Vermont	90	357	282	40	22	115
Central Atlantic (PADD 1B)	12,921	15,306	10,429	9,111	13,076	8,293
Delaware	99	88	122	75	168	70
District of Columbia	598	291	165	265	693	300
Maryland	3,950	4,772	2,807	3,249	3,646	1,162
New Jersey	3,951	4,562	2,331	2,127	2,408	3,323
New York	3,058	3,656	3,288	2,460	2,394	785
Pennsylvania	1,265	1,939	1,715	936	3,767	2,654
Lower Atlantic (PADD 1C)	22,547	26,669	34,106	30,304	39,564	24,342
Florida	4,444	6,115	4,370	5,481	6,323	6,043
Georgia	2,502	2,936	3,644	3,282	3,357	2,957
North Carolina	3,848	5,265	18,456	16,839	12,133	2,794
South Carolina	6,479	8,859	4,029	1,911	1,436	636
Virginia	4,812	3,460	3,259	2,611	16,167	11,630
West Virginia	462	33	347	179	148	281
Midwest (PADD 2)	7,549	8,934	8,526	7,053	7,483	5,948
Illinois	266	316	329	728	1,415	239
Indiana	12	225	46	121	31	45
Iowa	38	37	50	50	154	43
Kansas	812	614	586	436	341	87
Kentucky	919	944	470	1,110	1,022	467
Michigan	1,184	2,147	889	267	807	1,270
Minnesota	292	454	257	167	225	322
Missouri	1,738	1,939	1,881	1,596	2,226	1,748
Nebraska	48	118	178	111	116	107
North Dakota	367	422	1,361	1,576	446	661
Ohio	72	707	1,186	325	268	192
Oklahoma	452	312	295	80	125	457
South Dakota	927	131	171	20	53	24
Tennessee	181	95	630	270	81	15
Wisconsin	240	473	197	194	173	273
Gulf Coast (PADD 3)	45,555	38,923	33,652	40,713	35,803	10,729
Alabama	1,369	793	2,014	2,203	2,124	1,649
Arkansas	284	421	283	447	458	329
Louisiana	11,319	940	1,619	3,892	2,900	1,648
Mississippi	722	105	769	845	1,280	1,726
New Mexico	839	548	582	306	859	572
Texas	31,023	36,115	28,385	33,020	28,183	4,805
Rocky Mountain (PADD 4)	950	1,230	1,063	2,267	2,537	663
Colorado	553	761	715	922	1,166	365
Idaho	247	23	40	17	9	0
Montana	145	424	130	109	8	1
Utah	6	23	96	917	1,097	101
Wyoming	0	0	81	303	257	197
West Coast (PADD 5)	105,090	148,447	122,525	122,342	119,723	73,457
Alaska	6,835	7,305	7,875	6,369	7,234	5,748
Arizona	1,237	1,611	1,291	1,649	330	284
California	11,984	10,093	16,107	56,828	36,038	44,068
Hawaii	75,189	1,940	8,939	48,583	2,141	1,557
Nevada	1,868	1,560	1,684	1,745	1,983	1,205
Oregon	2,027	1,948	2,198	2,388	1,469	1,940
Washington	5,950	123,989	84,431	4,781	70,527	18,657

In thousand gallons. Data source: EIA 2013b.

End Notes

[1] The EPA considers co-processing to occur if both petroleum and biomass feedstock are processed in the same unit simultaneously.

[2] EIA. (2013). "Petroleum and Other Liquids - Definitions, Sources and Explanatory Notes," Accessed August 2011: http://www.eia.gov/dnav/pet/TblDefs/pet_cons_821dst_tbldef2.asp.

[3] In July 2011, President Barack Obama proposed that the CAFE standards for cars and light trucks increase to 54.5 miles per gallon (mpg) by 2025 (currently 30.2 mpg for passenger cars and 24.1 mpg for light trucks), the biggest increase since the federal government started regulating fuel economy in the 1970s. The auto industry will be given time to adapt to the proposed 2025 standard. Under the supplemental notice of intent, passenger cars are required to increase fuel economy from 2017-2021 by 4.1% annually while light-duty trucks are required to have 2.9% annual improvements.

[4] EIA. (2013). "Petroleum and Other Liquids - Definitions, Sources and Explanatory Notes," Accessed August 2011: http://www.eia.gov/dnav/pet/TblDefs/pet_cons_821dst_tbldef2.asp.

[5] These residues include unutilized wood volume from cut--or otherwise killed--growing stock from cultural operations, such as pre-commercial thinnings or from timberland clearing. This category does not include volume removed from inventory through reclassification of timber land to productive reserved forest land.

[6] Other forestlands are defined as incapable of producing at least 20 cubic feet per acre per year of industrial wood under natural conditions because of a variety of adverse site conditions including poor soils, lack of rainfall, and high elevation.

[7] Primary mill residues include wood materials (coarse and fine) and bark generated at manufacturing plants (primary wood-using mills) when round wood products are processed into primary wood products like slabs, edgings, trimmings, sawdust, veneer clippings and cores, and pulp screenings (USDA – Forest Service). Secondary mill residues include wood scraps and sawdust from woodworking shops such as furniture factories, wood container and pallet mills, wholesale lumberyards, and flooring.

[8] This category includes wood that has a commercial value for other uses but is used as an energy feedstock because of competitive market conditions.

[9] This volume includes logging residues and other removals, unused primary mill residues, secondary mill residues, and urban wood waste.

[10] NAS (2009) evaluates corn/wheat residues and hay only, while Milbrandt (2005) considers residues from 18 major agricultural crops.

[11] In 2013, biomass-based diesel mandate was revised upwards from 1 billion gallons to 1.28 billion gallons actual volume (EPA 2013b).

[12] West Texas Intermediate crude oil price: $100.54/barrel.

INDEX

A

B

C

D

E

F

J

K

L

M

N

O

P

Q

R

S

T

U

V

W

Y